Elements of Ordinary Differential Equations and Special Functions

A. Chakrabarti
Department of Mathematics
Indian Institute of Science
Bangalore, India

[SECOND EDITION]

PUBLISHING FOR ONE WORLD

NEW AGE INTERNATIONAL (P) LIMITED, PUBLISHERS
New Delhi • Bangalore • Chennai • Cochin • Guwahati • Hyderabad
Jalandhar • Kolkata • Lucknow • Mumbai • Ranchi
Visit us at **www.newagepublishers.com**

Published by New Age International (P) Ltd., Publishers
First Edition: 1990
Second Edition: 1996
Reprint: 2008

Branches:

- 36, Malikarjuna Temple Street, Opp. ICWA, Basavanagudi, **Bangalore**. © (080) 26677815
- 26, Damodaran Street, T. Nagar, **Chennai**. © (044) 24353401
- Hemsen Complex, Mohd. Shah Road, Paltan Bazar, Near Starline Hotel, **Guwahati**. © (0361) 2543669
- No. 105, 1st Floor, Madhiray Kaveri Tower, 3-2-19, Azam Jahi Road, Nimboliadda, **Hyderabad**. © (040) 24652456
- RDB Chambers (Formerly Lotus Cinema) 106A, Ist Floor, S.N. Banerjee Road, **Kolkata**. © (033) 22275247
- 18, Madan Mohan Malviya Marg, **Lucknow**. © (0522) 2209578
- 142C, Victor House, Ground Floor, N.M. Joshi Marg, Lower Parel, **Mumbai**. © (022) 24927869
- 22, Golden House, Daryaganj, **New Delhi**. © (011) 23262370, 23262368

ISBN (10): 81-224-0880-X
ISBN (13): 978-81-224-0880- 5

Rs. 160.00

C-08-10-3190

Printed in India at Nisha Enterprises, Delhi.

PUBLISHING FOR ONE WORLD
NEW AGE INTERNATIONAL (P) LIMITED, PUBLISHERS
4835/24, Ansari Road, Daryaganj, New Delhi-110002
Visit us at **www.newagepublishers.com**

In Memory of
My Mother

Preface to the Second Edition

In the second edition of this book, efforts have been made to correct the printing errors that existed in the first edition, and, for which an errata list was included in that edition, which might have caused certain disadvantages to the readers.

Also in this new edition there has been added a new chapter (Chapter-10) entitled "The Green's function technique". Like in all the other chapters of the book, in this new Chapter 10, this very powerful technique of solving Linear Ordinary Differential Equations has been dealt with in as short and lucid a manner as has been possible. It is hoped that by the addition of this new Chapter, a major portion of the theory of ordinary differential equations has also been taken care of and it will, therefore, serve many important purposes for users working in Applied Mathematics and Theoretical Physics.

A. Chakrabarti

Preface to the First Edition

The theory of ordinary differential equations and special functions constitutes a vast area of Mathematics, and the aim of this book is just to bring out the essential elementary feature of this subject. The various ideas and results reported in the present book are expected to be helpful to the students of Applied Mathematics, Physics and Engineering.

The book is based upon a course of lectures delivered to the students of Bachelor of Engineering at the Indian Institute of Science, Bangalore, India, for a number of years starting 1968. All these years, the students have expressed their sincere desire to have a book of this type containing the essential basic informations in a single volume which are otherwise available in a scattered form, in several books, written on this subject, by several authors.

It is for the benefit of the beginners that I have tried to be as lucid as possible and have included several worked out problems so that the main theme of the subject can be followed without much difficulty.

The topic "Series solution near a Regular Singular point" tackled in Chapter IV based on the wellknown Frobenius' method should constitute the central part of the book, as the studies on several special functions, such as the Legendre and the Bessel functions, tackled in Chapter V and VI, rely heavily on the methods described in Chapter IV.

The topic of existence and uniqueness of solution of Ordinary Differential Equations has been deferred till Chapter VIII, in order to create motivation of studying this topic by way of introducing the students to the general subject first, and its theory, next.

Chapter IX, on "Eigenvalue Problems" has been presented in a very short form, exposing the reader only to the elementary concepts of this large class of problems associated with Ordinary Differential Equations.

An appendix to the book is provided at the end, in order to acquaint the readers with various other special functions, such as the Gamma function, the Beta functions, the Hypergeometric functions and the Chebyshev polynomials, which have not been discussed in the main text, in order to make the book as short as possible, but at the same time to show the importance of these special functions also.

I have been helped very much by the monographs of E.A. Coddington (3), E.L. Ince (6), I.N. Sneddon (12), W.W. Bell (1), N.N. Lebedev (8), and M.S.P.

Eastham (4), on this subject, and also the monograph of W.E. Williams (16) on "Partial Differential Equations". The maximum help that I have received has been the existence of a detailed lecture notes on this subject, prepared by the late Professor P.L. Bhatnagar who delivered his lectures before 1968, when I joined the Indian Institute of Science, Bangalore.

I wish to thank all my students who have raised interesting questions to me while teaching the course, all of which have been of immense help while writing this book.

I thank the Centre for Continuing Education at the Indian Institute of Science, for partial financial support regarding the preparation of the manuscript.

I am thankful to Mr. Y.S. Ramaswamy, who has elegantly typed the manuscript with lot of patience.

My special thanks are due to my wife Dr. (Mrs.) Sthiti Chakrabarti and my little son, Master Amit Chakrabarti who missed me very much for the last couple of years when I was engaged in writing this book, in addition to my other responsibilities in research.

A. Chakrabarti

Contents

CHAPTER - ONE

Basic Definitions and Properties

ORDINARY DIFFERENTIAL EQUATIONS occur, in a natural way, while studying certain problems of physics and engineering. The object of the present book is to describe an elementary theory of such differential equations, along with some important techniques of solving the equations of most frequent occurrence.

1.1 BASIS DEFINITIONS

We start with some basic definitions of the theory of Ordinary Differential Equations and explain briefly certain important properties of the solutions of such equations.

Definition 1.1.1

An equation of the form

$$F(x,\ y,\ y^{(1)},\ y^{(2)},\ \ldots,\ y^{(n)}) = 0 \tag{1.1}$$

where F is a given function (real or complex valued) of the variables occurring within the parenthesis, $y = y(x)$, is an unknown function of a real variable x, superscripts denoting derivatives with respect to (w.r.t) x, e.g. $y^{(1)} = \frac{dy}{dx}$, $y^{(4)} = \frac{d^4 y}{dx^4}$, is called an 'Ordinary Differential Equation' for the determination of the unknown function $y(x)$.

Definition 1.1.2

The positive number 'n' occurring in equation (1.1), representing the highest order of the derivative of y, is called the 'order' of the Ordinary Differential Equation (1.1).

Definition 1.1.3

If the function F in equation (1.1) is such that the equation can be expressed in the form

$$f_0(x)\, y^{(n)} + f_1(x)\, y^{(n-1)} + f_2(x)\, y^{(n-2)} + \ldots + f_{n-2}(x)\, y^{(2)} + f_{n-1}(x)\, y^{(1)} + f_n(x)\, y = Q(x), \tag{1.2}$$

where $f_i(x)$ $(i = 0, 1, 2, ..., n)$ and $Q(x)$ are known functions of x, then equation (1.1) is said to be 'Linear' and equations (1.1) which cannot be written in the form (1.2) are said to be 'Non-linear'.

While Linear Ordinary Differential Equations are of principal concern to us in the present book, some basic concepts concerning first order non-linear equations are described in Chapter 8, where a brief discussion on the Existence and Uniqueness theorem is also taken up.

1.2 LINEAR ORDINARY DIFFERENTIAL EQUATIONS

Following are some basic definitions.

Definition 1.2.1

If the right hand member $Q(x)$ of equation (1.2) is 'zero', then the equation is said to be 'homogeneous', and equations which are not homogeneous are called 'inhomogeneous'.

Definition 1.2.2

The values of x, for which the coefficient $f_0(x)$ of the highest order (n) derivative of y appearing in equation (1.2) vanishes, are called the 'singularities' or 'singular points' of the Linear Ordinary Differential Equation (1.2).

Example 1. (*Bessel Equation*)

$$x^2 y^{(2)} + xy^{(1)} + (x^2 - \nu^2)\, y = 0 \quad (\nu = a \text{ constant})$$

$x = 0$ is a 'singular point'.

Example 2. (*Legendre Equation*)

$$(1 - x^2)\, y^{(2)} - 2xy^{(1)} + n\,(n + 1)\, y = 0, \quad (n = a \text{ constant})$$

$x = \pm 1$ are 'singular points'.

Example 3. (*Hypergeometric Equation*)

$$x\,(1 - x)\, y^{(2)} + [\gamma - (\alpha + \beta + 1)\, x]\, y^{(1)} - \alpha\,\beta\, y = 0$$

$x = 0, 1$ are 'singular points'.

Definition 1.2.3

Any function $y = f(x)$ which satisfies equation (1.2) is said to be 'a solution' of the differential equation (1.2). If $f(x)$ contains 'n' arbitrary constants $c_1, c_2, \ldots, c_n$ (i.e., $f(x) = \phi(x, c_1, c_2, \ldots, c_n)$), then the function $f(x)$ is said to e the 'general solution' of equation (1.2). On the other hand, if $f(x)$ does not contain any arbitrary constant, then $f(x)$ is said to be a 'particular solution' (or Particular Integral (P.I.) of equation (1.2)). Finally, if $Q(x) = 0$ in equation (1.2), then the general solution of the homogeneous equation (1.2) is said to be the 'Complementary Function' (C.F.) of equation (1.2).

We shall now describe some important properties associated with solutions of Linear Ordinary Differential Equations.

Property I—Superposability of Solutions

If $Q(x)$ in equation (1.2) is of the form

$$Q(x) = \sum_{i=1}^{m} Q_i(x), \tag{1.3}$$

and if $y = y_i(x)$ is a solution of the equation

$$\sum_{s=0}^{n} f_s(x)\, y^{(n-s)} = Q_i(x), \tag{1.4}$$

they $y = \sum_{i=1}^{m} y_i(x)$ is a solution of equation (1.2), with $Q(x)$ as given by equation (1.3). The proof of the property of superposability of solutions is straightforward and can be completed easily by using *Definition* 1.2.3.

Note: The above property is satisfied only by linear ordinary differential equations and not by non-linear equations, as can be observed through the following special non-linear equation of second order:

$$\frac{d^2 y}{dx^2} + a\frac{dy}{dx} + by^2 = g_1(x) + g_2(x),, \quad (b \neq 0) \tag{1.5}$$

If $y = y_i(x)$ $(i = 1, 2)$ are the solutions of the equations

$$\frac{d^2 y}{dx^2} + a\frac{dy}{dx} + by^2 = g_i(x), (i = 1, 2), \quad (g_i \neq 0) \tag{1.6}$$

then by direct substitution of $y = y_1 + y_2$ in equation (1.5) gives that

$$2by_1 y_2 = 0, \tag{1.7}$$

which is not true unless y_1 or y_2 or both vanish identically, showing thereby, the failure of the property of superposability of solutions of the nonlinear equation (1.5)

Property II—Removal of the right hand member

Assuming that $y_1(x)$ is a particular solution of equation (1.2), and substituting $y = y_1(x) + u(x)$ there, we get

$$\sum_{s=0}^{n} f_s(x)\, u^{(n-s)} = 0, \tag{1.8}$$

and this is the homogeneous part of the differential equation (1.2), having its general solution (C.F.) of the form $u(x) = \psi(x, \alpha_1, \alpha_2, \ldots, \alpha_n)$, where $\alpha_1, \alpha_2, \ldots, \alpha_n$ are n arbitrary constants. We thus arrive at the following important property about the general solution of the Linear Ordinary Differential equation (1.2):

If (G.S.) = General Solution of eqn. (1.2), then

(G.S.) = (C.F.) + (P.I.) (See Definition 1.2.3).

Before taking up the next property, we shall define a different concept which is very useful in obtaining the general solution of Linear Ordinary Differential Equations.

Definition 1.2.4

A set of functions $(y_1(x), y_2(x), \ldots, y_n(x)) = \{y_i(x)\}_{i=1}^{n}$ is said to be 'linearly dependent' if there exist constants $\lambda_1, \lambda_2, \ldots \lambda_n$, not all of which are zero simultaneously, such that

$$\lambda_1 y_1 + \lambda_2 y_2 + \ldots + \lambda_n y_n = 0. \tag{1.9}$$

The immediate consequence of linear dependence of a given set of functions $\{y_i(x)\}_{i=1}^{n}$ is that, after assuming $\lambda_n \neq 0$ in equation (1.9), we obtain

$$y_n(x) = \sum_{i=1}^{n-1} \beta_i y_i \left(\beta_i = -\frac{\lambda_i}{\lambda_n} \right) \tag{1.10}$$

showing, therefore, that the function $y_n(x)$ can be expressed as a linear combination of the other $(n-1)$ functions $y_1, y_2, \ldots y_{n-1}$ of the given set.

Definition 1.2.5

If a set of functions $\{y_1(x)\}_{i=1}^{n}$ is not 'linearly dependent', then it is said to be 'linearly independent'.

Following are some typical examples of sets of linearly independent functions:

(i) $y_1(x) = 1, y_2(x) = x, y_3(x) = x^2, (n = 3)$

(ii) $y_i(x) = \exp(a_i x), (i = 1, 2, \ldots, n)\ (a_i \neq a_j, i \neq j)$

(a_i's are constants).

(iii) $y_1(x) = 1, y_2(x) = \cos x, y_3(x) = \cos 2x, \ldots$

$y_n(x) = \cos(n-1)x$.

Though a direct application of *Definition* 1.2.4 does not help establishing linear independence of a given set of functions, the following theorem, valid for sets of differentiable (upto order $(n-1)$ at least) functions, is extremely useful in the theory of Ordinary Differential equations.

Theorem 1.2.1

The necessary and sufficient conditions, for a set of diferentiable functions $\{y_i(x)\}_{i=1}^{n}$ to be linearly independent, is that the following determinant (known as the "Wronskian" of the set):

$$W(y_1, y_2, \ldots, y_n) = \begin{vmatrix} y_1 & y_2 & \ldots & y_n \\ y'_1 & y'_2 & \ldots & y'_n \\ \ldots\ldots & \ldots\ldots & \ldots & \ldots\ldots \\ y_1^{(n-1)} & y_2^{(n-1)} & \ldots & y_n^{(n-1)} \end{vmatrix}$$

is not 'identically equal to zero' in the 'domain of definition' of the given set of functions.

The proof of the theorem is immediate if one uses the definition 1.2.4 and 1.2.5 straightaway, and the details are left to the reader.

The above theorem gives the criterion that

$$W(y_1, y_2, \ldots, y_n) \not\equiv 0 \tag{1.11}$$

for 'linear independence' of a given set of functions and this criterion is directly applicable to each of the sets of functions given in the above example (i)-(iii).

We are now in a position to describe the following important property involving the solutions of the Homogeneous Linear Ordinary Differential Equation (1.8).

Property III. If $u_1(x), u_2(x), \ldots, u_n(x)$ are n linearly independent solutions of the homogeneous linear ordinary differential equation (1.8), then the linear combination

$$u(x) = \sum_{i=1}^{n} c_i\, u_i(x), \tag{1.12}$$

where c_i's are arbitrary constants, none of which is zero, is the 'general solution' of equation (1.8). That the expression on the right of equation (1.12) is a solution of the differential equation (1.8) is obvious and that the solution (1.12) is the general solution also follows easily by using definition 1.2.3.

Note: The general solution of the homogeneous equation (1.8) which is the same as the C.F. of equation (1.2) can, therefore, be written down in the form (1.12), as soon as 'n' 'linearly independent' solutions of equation (1.8) are determined. It is important to note at this stage, that the set of solutions $\{u_i(x)\}_{i=1}^{n}$ of equation (1.8) must be 'linearly independent' in order that (1.12) represents the general solution of (1.8). For, if otherwise, constants $\alpha_1, \alpha_2, \ldots, \alpha_n$ exist, all of which are not zero, such that $\alpha_1 u_1 + \alpha_2 u_2 + \ldots + \alpha_n u_n = 0$, and then, assuming $\alpha_n \neq 0$, we get

$$u_n = -\sum_{i=1}^{n-1} \frac{\alpha_i}{\alpha_n} u_i, \tag{1.13}$$

so that (1.12) gives that

$$u = \sum_{i=1}^{n-1} \beta_i u_i, \tag{1.14}$$

with $\beta_i = \left(c_i - \alpha_i \dfrac{c_n}{\alpha_n} \right)$ $(i = 1, 2, \ldots, n-1)$, giving rise to the solution (1.14) involving only $n-1$ arbitrary constants β_i, and therefore, the form (1.12) (same as (1.14)) under such circumstances ($\{u_i\}_{i=1}^{n}$ linearly dependent) does not represent the general solution of equation (1.8).

We next define an important concept, known as 'Fundamental solutions', of Linear Ordinary Differential Equations. To this end, we assume that our problem is to solve equation (1.8) under the subsidiary conditions

$$u = u(a),\ u^{(1)} = u^{(1)}(a),\ \ldots,\ u^{(n-1)} = u^{n-1}(a), \tag{1.15}$$

where 'a' is a fixed point in the 'domain of definition' of function $y(x)$.

In order to arrive at the solution of the problem posed as above, we can first write down the 'general solution' of equation (1.8) in the form

$$u(x) = \alpha_1 u_1 + \alpha_2 u_2 + \ldots + \alpha_n u_n, \tag{1.16}$$

where $\{u_i\}_{i=1}^{n}$ is a set of linearly independent solutions of (1.8), and then determine the n constant $\alpha_1, \alpha_2, \ldots, \alpha_n$ by asking the solution (1.16) to satisfy the 'n' conditions (1.15). But, as is obvious, such a method of determination of the constants α_i is usually very laborious and a lot of simplification can be effected by seeking the n 'linearly independent' solutions $u_i(x)$ to satisfy the following simple conditions.

$$\left.\begin{array}{lllll} u_1(a) = 1, & u_1^{(1)}(a) = 0, & u_1^{(2)}(a) = 0, & \ldots, & u_1^{(n-1)}(a) = 0, \\ u_2(a) = 0, & u_2^{(1)}(a) = 1, & u_2^{(2)}(a) = 0, & \ldots, & u_2^{(n-1)}(a) = 0, \\ \ldots\ldots & \ldots\ldots & \ldots\ldots & \ldots, & \ldots\ldots \\ u_n(a) = 0, & u_n^{(1)}(a) = 0, & u_n^{(2)}(a) = 0, & \ldots, & u_n^{(n-1)}(a) = 1. \end{array}\right\} \tag{1.17}$$

Using these special solutions $u_i(x)$, satisfying the conditions (1.17), in the general solutions (1.16), we thus observe that the conditions (1.15) will be satisfied if

$$\alpha_1 = u(a),\ \alpha_2 = u^{(1)}(a),\ \alpha_3 = u^{(2)}(a),\ \ldots,\ \alpha_n = u^{(n-1)}(a), \tag{1.18}$$

and then, the solution of our main problem posed above is obtained in the form:

$$u(x) = u(a)\,u_1(x) + u^{(1)}(a)\,u_2(x) + u^{(2)}(a)\,u_3(x) + \ldots + u^{(n-1)}(a)\,u_n(x), \tag{1.19}$$

We then define:

Definition 1.2.6.

The set of linearly independent solutions $\{u_i(x)\}_{i=1}^{n}$ of the homogeneous equation (1.8) is said to be a set of 'Fundamental solutions' if $u_i(x)$ satisfy the conditions (1.17).

We will be able to consider specific examples on 'Fundamental Solutions' afterwards, when we have learnt various methods of solutions of linear ordinary differential equations, which will be described in latter chapters.

PROBLEMS

1. Verify that x^3 and x^{-2} are solutions of the equation

$$\frac{d^2 y}{dx^2} - \frac{6}{x^2} y = 0.$$

Are these solutions linearly independent?

2. Verify that x is a particular solution of the equation

$$\frac{d^2 y}{dx^2} + y = x.$$

3. Let $u_1, u_2, \ldots, u_n$ be n linearly independent solutions of the equation

$$L(u) = p_0(x)\, u^{(n)} + p_1(x)\, u^{(n-1)} + \ldots + p_n(x)\, u = 0 \; (*)$$

Show that $W(u_1, u_2, \ldots, u_n) = W_0 \exp\left\{-\int_{x_0}^{x} \frac{p_1}{p_0}\, dx\right\}$,

where W_0 is the value of W when $x = x_0$, a fixed point.

Hints : Obtain dW/dx and use the identity

$$p_0\, u_r^{(n)} = -\sum_{i=1}^{n} p_i\, u_r^{(n-i)}, \quad (r = 1, 2, \ldots, n)$$

since each of u_r $(r = 1, 2, \ldots, n)$ is a solution of the equation (*). Get $\frac{dW}{dx} = -\frac{p_1}{p_0} W$ and integrate.

4. Verify that $y = c_1 \left(\frac{x-1}{x+1}\right)^{\alpha} + c_2 \left(\frac{x-1}{x+1}\right)^{\alpha'}$ is the general solution of the equation

$$(x^2 - 1)^2\, y^{(2)} + 2(x-1)(x - \alpha - \alpha')\, y^{(1)} + 4\,\alpha\,\alpha'\, y = 0,$$
$$(-1 < x < 1)$$

with c_1 and c_2 as arbitrary constants.

5. Verify that each of the functions

$$y_1(x) = 2e^{-x} - e^{-2x} \text{ and } y_2(x) = \frac{e^{-x} - e^{-2x}}{e^{-a} - e^{-2a}}$$

are solutions of the equation $\frac{d^2 y}{dx^2} + 3\frac{dy}{dx} + 2y = 0.$

Also verify that (i) $y_1 = 1$ and $dy_1/dx = 0$, at $x = 0$,

and that (ii) $y_2 = 0$ at $x = 0$ and $y_2 = 1$, at $x = a\ (> 0)$.

Note : (I) Problems of the type (i) associated with an ODE, where the unknown function and its derivatives or a combination of both satisfy given conditions at a single fixed point in the interval in which the ODE holds, are called "Initial Value Problems."

(II) Problems of the type (ii), when the unknown function and its derivatives or a combination of both satisfy given conditions at two end points of the interval of validity of the given ODE, are called 'Boundary Value Problems'.

CHAPTER - TWO

Special Methods of Solution of Ordinary Differential Equations

This chapter is devoted to some of the special methods of solution of Linear Ordinary Differential Equations. While the methods described for the solution of equations with constant coefficients are sufficiently general, in the sense that these methods are applicable to any such equation, the methods described for the solution of equations with variable coefficients are applicable only to specific problems. However, such special methods of solving Ordinary Differential Equations are immensely helpful in a number of situations of practical importance.

2.1 EQUATIONS WITH CONSTANT COEFFICIENTS

Writing $f^{(k)}(x) = \dfrac{d^k f}{dx^k} = D^k f$, $(k = 0, 1, 2, \ldots)$ with $D^0 f = f(x)$, the general n-th order linear ordinary differential equation with constant coefficients can be written in the symbolic form:

$$F(D)\, y = Q(x), \tag{2.1}$$

where

$$F(D)\, y = (a_0 D^n + a_1 D^{n-1} + a_2 D^{n-2} + \ldots + a_n)\, y, \tag{2.2}$$

and the right hand member $Q(x)$, is a known function with a_i's $(i = 0, 1, 2, \ldots, n)$ being known constants.

By using the concept of (C.F.) and (P.I.) as described in *definition* 1.2.3 of Chapter 1, we shall express the 'general solution' of equation (2.1) in the form

$$y_{GS} = y_{CF} + y_{PI}, \tag{2.3}$$

where y_{CF} represents the complementary function satisfying the homogeneous part of equation (2.1) and y_{PI} represents its particular solution. In what follows next, we shall describe a constructive method of arriving at the function y_{CF} first and take up, later on, the problem of determining the function y_{PI} in certain special circumstances, as the problem of determination of y_{PI}, in general, is by no means an easy task even though the method of variation of parameters (also discussed here) provide a useful idea in many interesting situations.

2.1.1 The Complementary Function (y_{CF})

Using the 'Fundamental Theorem of Algebra' (see Churchil [2]) we can express, the 'symbolic, expression $F(D)$, in the form

$$F(D) = a_0 (D - \alpha_1)(D - \alpha_2) \ldots (D - \alpha_n), \tag{2.4}$$

where $\alpha_1, \alpha_2, \ldots, \alpha_n$ are distinct or non-distinct real or complex constants representing the n roots of the polynomial equation

$$P(m) = 0, \quad (m = \text{unknown}) \tag{2.5}$$

Definition 2.1.1

The equation $F(m) = 0$ corresponding to the ordinary differential equation (2.1) is called its 'Auxilary Equation'.

Before proceeding further we shall prove the following important results associated with the 'symbolic operators' $F(D)$.

Result-I

$$F(D)(Ve^{ax}) = e^{ax} F(D + a) V,$$

where V is a 'differentiable, function, and 'a' is a constant, the derivative of V existing upto order n atleast.

Proof

Using Leibnitz's rule on differentiation of a product of two functions, we get, in usual notations,

$$D^r (Ve^{ax}) = e^{ax} D^r V + r_{c_1} a e^{ax} D^{r-1} V + r_{c_2} a^2 e^{ax} D^{r-2} V + \ldots + a^r e^{ax} V$$

$$= e^{ax} (D + a)^r V. \tag{2.6}$$

Then a straightforward utility of the results (2.6) proves Result I immediately. In equation (2.6) the quantities p_{c_q} represent the Binomial coefficients as defined by

$$p_{c_q} = \frac{p\,!}{q\,!\,(p - q)\,!} \quad (p > q) \tag{2.7}$$

where the factorial function $p\,!$ is given by $p\,! = p\,(p - 1) \ldots 3.2.1.$

Result-II

$$(D - a) y = 0 \Leftrightarrow y = Ae^{ax},$$

where A is an arbitrary constant, and a is a known constant.

Proof

The equation $(D - a)\, y = 0$ is the same as the equation

$$\frac{dy}{dx} - ay = 0. \tag{2.8}$$

and a straightforward integration gives

$$y = Ae^{ax}, \tag{2.9}$$

where A is an arbitrary constant.

Alternatively, substituting (2.9) in the left of equation (2.8) shows that the latter equation is satisfied, proving the indirect result.

Result-III

$$(D - a)\,(D - b)\, f = (D - b)\,(D - a)\, f.$$

Proof

$$\text{L.H.S.} = (D - a)\left(\frac{df}{dx} - bf\right) = \frac{d^2 f}{dx^2} - (a + b)\frac{df}{dx} + abf.$$

$$= \text{R.H.S.}$$

Hence the result.

Using the above two results we can easily prove the following theorem:

Theorem 2.1.1

If $F(D) = a_0 \prod_{r=1}^{n} (D - \alpha_r)$, in which all the α_r's are distinct, then the general solution of the equation

$$F(D)\, y = 0$$

is given by

$$y_{GS} = \sum_{r=1}^{n} A_r \exp(a_r x), \tag{2.10}$$

where A_r's are arbitrary constants.

Note : The expression on the right of equation (2.10) also represents y_{CF}. (See equation (2.3)) associated with the ordinary differential equation (2.1).

Result-IV

$$(D - a)^k\, y = 0 \Leftrightarrow y = \sum_{r=0}^{k-1} A_r x^r e^{ax}, \quad (k > 0, \text{integer})$$

where A_r's are arbitrary constants.

Proof

Writing $(D - a)^k\, y = (D - a)^k\, [e^{ax}\,(e^{-ax}\, y)]$, and using Result I, we get

$$(D-a)^k\, y = e^{ax}\,(D-a+a)^k\,(e^{-ax}\, y).$$

Hence the equation $(D-a)^k\, y = 0$ reduces to

$$D^k\,(e^{-ax}\, y) = 0. \tag{2.11}$$

from which, by repeated integration, the direct result follows quickly.

The indirect result can be verified by direct substitution. We can, therefore, prove the following important theorem which is very useful in writing down y_{CF} associated with equation (2.1).

Theorem 2.1.2

If $F(D) = a_0 \sum\limits_{r=1}^{p} (D-\alpha_r)^{n_r}$, $(n_1 + n_2 + \ldots, + n_p = n)$ then the general solution of $F(D)\, y = 0$ is given by

$$y_{GS} = \sum_{r=1}^{n} \sum_{s=0}^{n_r-1} A_{sr}\, x^s \exp(\alpha_r x), \tag{2.12}$$

where A_{sr} are arbitrary constants.

2.1.2 The Particular Integral (y_{PI})

While the two theorems 2.1.1 and 2.1.2 provide methods for obtaining y_{CF} for the general linear ordinary differential equation (2.1), with constant coefficients, the methods for obtaining y_{PI} of the same equation are not easy, in general, and we describe below certain special circumstances in which y_{PI} of equation (2.1) can be determined in a routine manner.

Case 1

$$Q(x) = x^m \qquad (m = a \text{ positive integer}).$$

We write

$$y_{PI} = \frac{1}{F(D)}\, x^m, \tag{2.13}$$

in a 'symbolic' manner, the meaning of which is that

$$F(D)\,(y_{PI}) = x^m. \tag{2.14}$$

If, in particular, $F(D) = D - a$, then the particular solution of (2.14) is given by

$$y_{FI} = e^{ax} \int e^{-ax}\, x^m\, dx,$$

which, by repeated integration by parts, gives

$$y_{PI} = -\frac{1}{a}\, x^m - \frac{m}{a^2}\, x^{m-1} - \frac{m(m-1)}{a^3}\, x^{m-2} - \ldots - \frac{m!}{a^{m+1}}$$

$$= -\frac{1}{a}\left(x^m + \frac{m}{a}x^{m-1} + \frac{m(m-1)}{a^2}x^{m-2} + \ldots + \frac{m!}{a^m}\right)$$

and the right hand side can be expressed in the form

$$-\frac{1}{a}\left(1-\frac{D}{a}\right)^{-1} x^m,$$

in a 'symbolic' manner, after expanding the expression $\left(1-\frac{D}{a}\right)^{-1}$ binomially and noticing that

$$D^p x^m = 0,$$

for $p > m$, and $D^m x^m = m!$.

We thus find, in a particular case when $F(D) = (D - a)$, that the y_{PI} satisfying the equation

$$F(D)(y_{PI}) = x^m,$$

is given by

$$y_{PI} = -\frac{1}{a}\left(1-\frac{D}{a}\right)^{-1} x^m, \tag{2.15}$$

symbolically speaking. This particular result (2.15), along with Result III proved earlier, provide a useful method of obtaining y_{PI} for the general equation (2.1) when its right hand member $Q(x)$ is a polynomial of a known degree, as illustrated in the following example :

Example 2.1.1

Obtain y_{PI} of the equation

$$F(D)\,y = (D^4 - 16)\,y = x^2 + 2x^4 - x^5.$$

Solution

We have

$$D^4 - 16 = (D-2)(D+2)(D-2i)(D+2i) \text{ with } i^2 = -1,$$

$$y_{PI} = \frac{1}{(D-2i)(D+2i)(D+2)}\left(-\frac{1}{2}\right)\left(1-\frac{D}{2}\right)^{-1}(x^2+2x^4-x^5)$$

$$= \frac{1}{(D-2i)(D+2i)(D+2)}\left(-\frac{1}{2}\right)\left(1+\frac{D}{2}+\frac{D^2}{4}+\frac{D^3}{8}+\frac{D^4}{16}+\frac{D^5}{32}\right)(x^2+2x^4-x^5)$$

or $$y_{PI} = \frac{1}{(D-2i)(D+2i)(D+2)}\left(-\frac{1}{2}\right)\left[x^2+x+\frac{1}{2}+2x^4+4x^3+6x^2 + 3x + 3 - x^5 - \frac{5}{6}x^4 - 5x^3 - \frac{15}{2}x^2 - \frac{15}{2}x - \frac{15}{4}\right] \tag{2.16}$$

and the other factors, involving $(D+2)$, $(D+2i)$ and $(D-2i)$ can be dealt with in a similar fashion.

It so happens that in this particular problem, and in all problems in general, (Proof is left to the reader), in which $Q(x)$ is a polynomial, we can derive y_{PI} much more quickly, in the following manner.

We write

$$y_{PI} = \frac{1}{(D^4-16)}(x^2+2x^4-x^5)$$

$$= -\frac{1}{16}\left(1-\frac{D^4}{16}\right)^{-1}(x^2+2x^4-x^5)$$

$$= -\frac{1}{16}\left[1+\frac{D^4}{16}\right](x^2+2x^4-x^5) \qquad (2.17)$$

after neglecting the terms involving higher powers of D in a binomial expansion of $\left(1-\dfrac{D^4}{16}\right)^{-1}$, taking note of the degree of the polynomial on the right of the operator.

Finally, equation (2.17) gives

$$y_{PI} = -\frac{1}{16}\left[x^2+2x^4-x^5+\frac{3}{2}-\frac{15}{2}x\right]$$

which is the same as the one derivable from (2.16).

Case II

$Q(x) = V(x)\,e^{ax}$, where $V(x)$ is a differentiable function and 'a' is a constant.

Writing, in this case also,

$$y_{PI} = \frac{1}{F(D)}\{V(x)\,e^{ax}\},$$

we get

$$F(D)\{y_{PI}\} = V(x)\,e^{ax},$$

which can be written as

$$F(D)\,[e^{ax}\{e^{-ax}\,y_{PI}\}] = V(x)\,e^{ax},$$

and this, by using Result I, takes up the form

$$e^{ax}F(D+a)\,[e^{-ax}\,y_{PI}] = V(x)\,e^{ax},$$

$$e^{-ax}\,y_{PI} = \frac{1}{F(D+a)}V(x)$$

$$\Rightarrow \qquad y_{PI} = e^{ax} \frac{1}{F(D+a)} V(x). \tag{2.18}$$

The problem of determination of y_{PI} thus is reduced to that of computing the expression $\frac{1}{F(D+a)} V(x)$ and this can be achieved fairly easily by the method described in Case I, when $V(x)$ is a polynomial in x, as illustrated below:

Example 2.1.2

Find y_{PI} of the equation

$$F(D)\, y \equiv (D^3 - 1)\, y = x^3 e^x$$

Solution

$$y_{PI} = \frac{1}{(D^3-1)} (x^3 e^x)$$

$$= e^x \frac{1}{(D+1)^3 - 1} x^3$$

$$= e^x \frac{1}{(D^3 + 3D^2 + 3D)} x^3$$

$$= e^x \frac{1}{D^2 + 3D + 3} (x^4/4), \left[\text{since } \frac{1}{D} x^3 = \frac{x^4}{4}, \text{ since } D(x^4/4) = x^3.\right]$$

we thus obtain

$$y_{PI} = \frac{e^x}{4} \frac{1}{(D-\alpha_1)(D-\alpha_2)} x^4,$$

with $\alpha_{1,2} = \frac{1}{2}(-3 \pm i\sqrt{3})$, and this can be handled by the method already described earlier, in *Case I.*

The above two cases are basic, as far as the problem of determination of y_{PI} for the ordinary differential equation (2.1) is concerned, in the sense that several other types of the right hand member of the equation, appearing in application, can be cast into one of the two types discussed.

For example, if $Q(x) = (\text{or } {}^{\sin}_{\cos} (ax))\, V(x)$, we can express $Q(x)$ in the alternative form:

$$Q(x) = \frac{V}{2i} (e^{iax} - e^{-iax}),$$

or

$$Q(x) = \frac{V}{2} (e^{iax} + e^{-iax}), \tag{2.19}$$

and proceed as in *Case II*.

If $Q(x)$ cannot be expressed in one of the forms described in the above two cases, we shall utilise the following important result.

Result-V

If $(D - a)\, y = f(x)$, then $y = e^{ax} \int e^{-ax} f(x)\, dx$ can be treated as a particular solution.

Proof

Using *Result I*, we get,

$$(D - a)\,[e^{+ax}\,(e^{-ax}\, y)] = f(x) \Rightarrow D\,(e^{-ax}\, y) = e^{-ax} f(x)$$

and a direct integration produces the result, after leaving out the constant of integration, as we are only interested in a particular solution.

The *Result V*, along with the *Result III* is immensely helpful in obtaining the 'particular integral' of the differential equation (2.1), i.e.,

$$F(D)\, y = Q(x),$$

after using the factorization (2.4), i.e.,

$$F(D) = a_0\,(D - \alpha_1)\,(D - \alpha_2) \ldots (D - \alpha_h)$$

in the case of any general form of $Q(x)$, as illustrated through the following example.

Example 2.1.3

Obtain y_{PI} for the equation

$$\frac{d^2 y}{dx^2} + a^2 y = \sec ax.$$

Solution

We first write the given equation in the form

$$(D - ia)\,(D + ia)\, y = \sec ax.$$

Then, using Result *V*, we get, as a particular solution,

$$(D + ia)\, y = e^{+iax} \int e^{-aix} \sec ax\, dx = e^{+iax} \left[x + \frac{1}{a} \log \cos (ax) \right]$$

which, in turn (use Result *V* again), gives

$$y\,(\equiv y_{PI}) = e^{-iax} \int e^{+2iax} \left(x + \frac{i}{a} \log \cos (ax) \right) dx,$$

and the final result can be derived by employing the rule of integration by parts.

The computation of y_{PI} for this particular example can also be simplified as follows:

Write

$$y_{PI} = \frac{1}{(D^2 + a^2)} \sec (ax) = \frac{1}{2ia}\left[\frac{1}{D - ia} - \frac{1}{D + ia}\right] \sec(ax)$$

$$= \frac{1}{2ia} [y_1 - y_2], \text{ (say)}$$

where y_1 and y_2 are the two particular solutions of the two equations

$$(D - ia)\, y_1 = \sec (ax) \tag{2.20}$$

and

$$(D + ia)\, y_2 = \sec (ax) \tag{2.21}$$

giving, by the help of *Result V*,

$$y_1 = e^{iax} = \int e^{-iax} \sec (ax)\, dx = e^{iax}\left[x + \frac{i}{a} \log (\cos\ ax)\right]$$

and

$$y_2 = e^{-iax}\left[x - \frac{i}{a} \log (\cos ax)\right],$$

so that we obtain

$$y_{PI} = \frac{x}{a} \sin (ax) + \frac{1}{a^2} \cos (ax) \log (\cos ax) \tag{2.22}$$

2.2 EQUATIONS OF EULER-CAUCHY TYPE

Definition 2.2.1

The linear Ordinary Differential equation

$$p_0\, x^n\, y^{(n)} + p_1\, x^{n-1}\, y^{(n-1)} + \ldots + p_n\, y = q\, (x) \tag{2.2.1}$$

in which $p_0, p_1, \ldots, p_n$ are known constants is called an equation of Euler-Cauchy type.

The methods of solution of equations with constant coefficients, as described in the previous section, are found to be applicable for equations of 'Euler-Cauchy' type (2.2.1), if the following transformations of the variables are utilized:

$$x = e^t,\ y\,(x) = Y\,(t),\ q\,(x) = Q\,(t), \tag{2.2.2}$$

giving

$$x\frac{dy}{dx} = \frac{dY}{dt},$$

$$x^2 \frac{d^2 y}{dx^2} = x \frac{d}{dx}\left(x \frac{dy}{dx}\right) - x \frac{dy}{dx} = \frac{d^2 Y}{dt^2} - \frac{dY}{dt} = \frac{d}{dt}\left(\frac{d}{dt} - 1\right) Y, \quad \text{etc.}$$

so that if we use the new differential operator $\theta = d/dt$, we obtain,

$$x D = \theta, \; x^2 D^2 y = \theta (\theta - 1) Y, \tag{2.2.3}$$

where $D = d/dx$ is the original differential operator appearing in equation (2.2.1).

We can also deduce that

$$x^n D^n y = \theta (\theta - 1) (\theta - 2) \ldots (\theta - n + 1) Y \tag{2.2.4}$$

for general values of n, and then, with the help of this result (2.2.4), equation (2.2.1) can be expressed in the form

$$\sum_{j=0}^{n} c_j \theta_j Y = Q(t), \tag{2.2.5}$$

where the constant coefficients c_j of the new differential equation (2.2.5) can be expressed, in terms of the constant p_j appearing in equation (2.2.1), in a straight-forward manner, as illustrated through the examples considered below.

We thus observe that the equations of Eluler-Cauchy type are reducible to equations with constant coefficients for which the methods of solution described in Section 2.1 are applicable directly.

Example 2.2.1

Obtain the general solution of

$$x^2 \frac{d^2 y}{dx^2} + 5x \frac{dy}{dx} + 4y = \cos(\log x).$$

Solution

Setting $x = e^t$, $y(x) = Y(t)$, $\theta = d/dt$, the given equation is reduced to the new equation

$$[\theta(\theta - 1) + 5\theta + 4] Y = \cos t,$$

or

$$(\theta^2 + 4\theta + 4) Y = \cos t.$$

or

$$(\theta + 2)^2 Y = \cos t.$$

We then get

$$Y_{CF} = (A_1 + A_2 t) e^{-2t}, \quad (A_1, A_2 = \text{Arbitrary constants})$$

and

$$Y_{PI} = \frac{1}{(\theta+2)^2}\frac{1}{2}[e^{it}+e^{-it}]$$

$$= \frac{1}{2}\left[e^{it}\frac{1}{(\theta+2+i)^2}1 + e^{-it}\frac{1}{(\theta+2-i)^2}1\right]$$

$$= \frac{1}{2}\left[\frac{e^{it}}{(2+i)^2}+\frac{e^{-it}}{(2-i)^2}\right]$$

$$= \frac{1}{50}[(2-i)^2 e^{it}+(2+i)^2 e^{-it}]$$

$$= \frac{1}{50}(6\cos t + 8\sin t).$$

Transferring back to the old variables, we finally obtain

$$y_{GS} = (A_1 + A_2\log x)\,x^{-2} + \frac{1}{25}[3\cos(\log x) + 8\sin(\log x)].$$

Example 2.2.2

Find the general solution of

$$(2x+1)^2 y^{(2)} + (2x+1)\,y^{(1)} + 3y = \cos^2[\log(2x+1)]$$

Solution

First write $2x + 1 = x'$, $y = y'$ and obtain

$$y^{(1)} = 2\frac{dy'}{dx'},\; y^{(2)} = 4\frac{d^2y'}{dx'^2},$$

so that the given equation can be cast into the form

$$4x'^2\frac{d^2y'}{dx'^2} + 2x'\frac{dy'}{dx'} + 3y' = \cos^2(\log x')$$

and this equation is of the Euler-Cauchy type.

The general solution can, therefore, be obtained systematically as in example 2.1.1, after replacing $\cos^2 z$ by its equivalent form

$$\frac{1}{4}(e^{2tz} + e^{-2tz} + 2).$$

2.3 SOME SPECIAL METHODS

We shall describe, in this section, certain special methods for solving linear ordinary differential equations of *second order*. Similar, methods are also applicable to higher order equations. These methods are extremely useful for equations with variable coefficients which will be discussed systematically in the next chapter.

2.3.1 Removal of first order derivatives

Consider the second order equation

$$\frac{d^2 y}{dx^2} + p(x) \frac{dy}{dx} + q(x) y = 0. \tag{2.3.1}$$

Let $y(x) = u(x) v(x)$ be the general solution of (2.3.1).

Then we must have

$$u \frac{d^2 v}{dx^2} + \left(2 \frac{du}{dx} + pu\right) \frac{dv}{dx} + \left(\frac{d^2 u}{dx^2} + p \frac{du}{dx} + qu\right) v = 0. \tag{2.3.2}$$

If we now select the function $u(x)$ in such a manner that

$$2 \frac{du}{dx} + pu = 0, \tag{2.3.3}$$

then equation (2.3.2) will be free from the term involving the first order derivative dv/dx of the other unknown function v.

We thus find that the problem of solving equation (2.3.1) gets reduced to that of solving a simpler equation (in many circumstances) as given by

$$u \frac{d^2 v}{dx^2} + \left(\frac{d^2 u}{dx^2} + p \frac{du}{dx} + qu\right) v = 0, \tag{2.3.4}$$

in which the function u is to be replaced by the solution of equation (2.3.3).

The above method is illustrated by the following example.

Example 2.3.1

Solve $y^{(2)} + 4xy^{(1)} + (4x^2 + 11) y = 0.$

Solution

Assuming $y = uv$, equation (2.3.3) takes the form

$$2 \frac{du}{dx} + 4xu = 0,$$

having the general solution

$$u = c_1 e^{-x^2}, \tag{i}$$

where c_l is an arbitrary constant.

Then obtain

$$\frac{du}{dx} = -2c_1 xe^{-x^2}, \quad \frac{d^2 u}{dx^2} = -2c_1 (1 - 2x^2) e^{-x^2},$$

and find that equation (2.3.4) takes up the form

$$c_1 e^{-x^2} \frac{d^2 v}{dx^2} + [-2c_1 (1 - 2x^2) - 8c_1 x^2 + c_1 (4x^2 + 11)] e^{-x^2} v = 0$$

or

$$\frac{d^2 v}{dx^2} + 9v = 0, \tag{ii}$$

after cancelling the factor $c_1 e^{-x^2}$ throughout. Equation (ii) is a simple equation and its general solution is given by

$$v = c_2 \cos 3x + c_3 \sin 3x \tag{iii}$$

where c_2 and c_3 are arbitrary constants.

The general solution of the given equation is thus obtained in the form

$$y_{GS} = e^{-x^2} (A \cos 3x + B \sin 3x), \tag{iv}$$

where A $(= c_1 c_2)$ and B $(= c_1 c_3)$ are only two arbitrary constants, as expected to be present in the form of the general solution of a second order ordinary differential equation.

2.3.2 Utility of One of the Solutions

If, by some means (usually guess), one of the two independent solutions of a second order linear homogeneous equation is known to us, we can find out the general solution of the equation by using a method as described below.

Let the given differential equation be

$$\frac{d^2 y}{dx^2} + p(x) \frac{dy}{dx} + q(x)\, y = 0, \tag{2.3.5}$$

and let a solution $y_1(x)$ of the equation be known.

Then, assume the general solution of equation (2.3.5) in the form

$$y(x) = u(x)\, y_1(x), \tag{2.3.6}$$

where the function $u(x)$ is to be determined.

We find that the form (2.3.6) is a solution of equation (2.3.5) only if $u(x)$ satisfies the following equation [obtained by direct substitution of (2.3.6) into (2.3.5)].

$$\frac{d^2 u}{dx^2} + \left[p(x) + \frac{2}{y_1} \frac{dy_1}{dx} \right] \frac{du}{dx} = 0. \tag{2.3.7}$$

The general solution of equation (2.3.7) can be easily derived in the form

$$u(x) = A \int \frac{\exp\left\{ -\int p(x)\, dx \right\}}{y_1^2}\, dx + B, \tag{2.3.8}$$

where A and B are arbitrary constants.

The general solution (2.3.6) thus gets completely determined, by using (2.3.8). The method is illustrated through the following example:

Example 2.3.2

Obtain the general solution of

$$(x-1)\frac{d^2 y}{dx^2} - x\frac{dy}{dx} + y = 0.$$

Solution

It can be easily verified that the function $y_1(x) = x$ is a solution of the given equation. Then the formula (2.3.8) gives

$$u(x) = B + A\int \frac{\exp\left\{\int \frac{x}{x-1}\,dx\right\}}{x^2}\,dx$$

$$= B + A\int \frac{(x-1)\,e^x}{x^2}\,dx$$

$$= B + A\int d\,(e^x/x) = B + A\frac{e^x}{x}.$$

Thus, the general solution is given by

$$y = u(x)\,y_1(x) = Bx + Ae^x.$$

It is interesting to verify that the solution $y = x$ can also be derived by using the function e^x as the known solution, playing the role of $y_1(x)$ in the formula (2.3.8).

A similar technique can be employed in solving higher order homogeneous differential equations as illustrated through the following example.

Example 2.3.3

Solve the equation

$$\frac{d^3 y}{dt^3} - \frac{d^2 y}{dt^2} - \frac{dy}{dt} + y = 0. \tag{1*}$$

Solution

It can be verified that $y_1 = e^t$ and $y_2 = e^{-t}$ are two independent solutions of the given equation.

We assume, then, the general solution in the form

$$y(t) = u_1(t)\,y_1(t) + u_2(t)\,y_2(t), \tag{2*}$$

where u_1 and u_2 are two unknown functions to be determined.

The relation (2*) gives that

$$\frac{dy}{dt} = u_1 e^t - u_2 e^{-t}, \tag{3*}$$

under the condition that

$$\frac{du_2}{dt} = - e^{2t} \frac{du_1}{dt}. \tag{4*}$$

We then get

$$\frac{d^2 y}{dt^2} = u_1 e^t + u_2 e^{-t} + 2e^t \frac{du_1}{dt},$$

and

$$\frac{d^3 y}{dt^3} = u_1 e^t - u_1 e^{-t} + 2e^t \frac{d^2 u_1}{dt^2} + 2e^t \frac{du_1}{dt}. \tag{5*}$$

Substituting the values of

$$\frac{dy}{dt}, \frac{d^2 y}{dt^2} \text{ and } \frac{d^3 y}{dt^3},$$

thus obtained, into the given differential equation, we obtain

$$\frac{d^2 u_1}{dt^2} = 0$$

giving

$$u_1 = c_1 t + c_2, \tag{6*}$$

where c_1 and c_2 are two arbitrary constants.

The relation (4*) then determines the function $u_2 (t)$ in the form

$$u_2 (t) = -\frac{c_1}{2} e^{2t} + c_3, \tag{7*}$$

where c_3 is a third arbitrary constant.

Going back to (2*), we thus obtain the general solution, in the form

$$y = c_2 e^t + c_1 te^t + c_3 e^{-t} - \frac{c_1}{2} e^t$$

$$= Ae^t + Be^{-t} + Cte^t. \tag{8*}$$

where A, B and C are three arbitrary constants, as expected.

Note : The problem in Example 2.3.3 above can be easily handled by the methods described in 2.1.1. The method employed above is just to show how a knowledge of less number of solutions, than what is required, can be exploited to obtain the general solution of higher order homogeneous linear ordinary differential equations.

2.4 THE METHOD OF VARIATION OF PARAMETERS

The method of variation of parameters is an extremely useful method of finding out the 'Particular Integral' of a given inhomogeneous linear differential equation, and it uses the knowledge of the 'Complementary Function' of the given equation, as described below.

Consider the second order equation

$$y^{(2)} + p(x)\, y^{(1)} + q(x)\, y = r(x) \qquad (2.4.1)$$

Let the complementary function of (2.4.1) be written as

$$y_{CF} = c_1\, y_1(x) + c_2\, y_2(x), \qquad (2.4.2)$$

where c_1 and c_2 are two arbitrary constants.

Then we vary the constants c_1 and c_2 after assuming them as arbitrary functions of x (hence the name 'variation of parameters') and write the 'general solution' of (2.4.1) in the form:

$$y = c_1(x)\, y_1(x) + c_2(x)\, y_2(x), \qquad (2.4.3)$$

where the functions $c_1(x)$ and $c_2(x)$ are determined as follows:

Step—1

Select $c_1(x)$ and $c_2(x)$ in such a manner that

$$y^{(1)} = c_1\, y_1^{(1)}(x) + c_2\, y_2^{(1)}(x), \qquad (2.4.4)$$

i.e.,

$$c_1^{(1)}\, y_1 + c_2^{(1)}\, y_2 = 0. \qquad (2.4.5)$$

Step—2

Using (2.4.4), we get

$$y^{(2)} = c_1\, y_1^{(2)} + c_2\, y_2^{(2)} + c_1^{(1)}\, y_1^{(1)} + c_2^{(1)}\, y_2^{(1)}. \qquad (2.4.6)$$

Step—3

Substitute (2.4.3), (2.4.4) and (2.4.6) into the given equation (2.4.1) and obtain

$$c_1\,[y_1^{(2)} + p(x)\, y_1^{(1)} + q(x)\, y_1] + c_2\,[y_2^{(2)} + p(x)\, y_2^{(1)} + q(x)\, y_2] + c_1^{(1)}\, y_1^{(1)} + c_2^{(1)}\, y_2^{(1)} = r(x) \qquad (2.4.7)$$

Step—4

Use the fact that y_1 and y_2 are two solutions of the homogeneous part of equation (2.4.1), in equation (2.4.7) and obtain

$$c_1^{(1)}\, y_1^{(1)} + c_2^{(1)}\, y_2^{(1)} = r(x). \qquad (2.4.8)$$

Step—5

Solve the two equations (2.4.5) and (2.4.8) for the two unknown $c_1^{(1)}$ and $c_2^{(1)}$ and obtain:

$$c_1^{(1)}(x) = -\frac{r(x)}{W(x)} y_2, \quad c_2^{(1)}(x) = \frac{r(x)}{W(x)} y_1, \tag{2.4.9}$$

where the function

$$W(x) = y_1 y_2^{(1)} - y_2 y_1^{(1)} \tag{2.4.10}$$

represents the non-zero 'Wronskian' of the linearly independent solutions y_l and y_2 of the homogeneous equation.

Step—6

Integrate the two first order equation in (2.4.9) and obtain

$$\left.\begin{aligned} c_1(x) &= A_1 - \int \frac{r(x)\, y_2(x)}{W(x)}\, dx \\ \text{and} \quad c_2(x) &= A_2 + \int \frac{r(x)\, y_1(x)}{W(x)}\, dx \end{aligned}\right\} \tag{2.4.11}$$

where A_1 and A_2 are two arbitrary constants.

The general solution (2.4.3) is now obtainable by using (2.4.11) and it is clearly visible that the C.F. and P.I. are separated out ultimately in the desired form.

The following example shows the power of the method, described above, in a natural way:

Example 2.4.1

Find the general solution of the equation

$$(x-1)\, y^{(2)} - xy^{(1)} + y = (x-1)^2.$$

Solution

From *example* 2.3.2, we find that

$$r(x) = x - 1, \; y_1 = x, \; y_2 = e^x.$$

Therefore,

$$W(x) = xe^x - e^x = e^x (x-1),$$

and the relations (2.4.11) give that

$$c_1(x) = A_1 - \int dx = A_1 - x$$

$$c_2(x) = A_2 + \int xe^{-x}\, dx = A_2 - (x+1)\, e^{-x}.$$

Hence,

$$y_{GS} = A_1 x + A_2 e^x - x^2 - (x+1).$$

Considering the fact that, for the given equation we already have

$$y_{CF} = A_1 x + A_2 e^x,$$

we find that

$$y_{PI} = -1 - x - x^2,$$

and this can be verified by means of direct substitution in the given equation.

2.5 SOME MORE METHODS AND RESULTS INVOLVING EQUATIONS WITH CONSTANT COEFFICIENTS

2.5.1 Simultaneous Linear Ordinary Differential Equations with Constant Coefficients

In order to solve a system of linear ordinary differential equations with constant coefficients, we can develop general methods by using the results derived in section 2.1 on polynomial operators $f(D)$ and using the analogy with systems of algebraic equations, treating D as a pure number, in a formal manner.

We illustrate the procedure through the following system of just three equations:

$$\left.\begin{aligned} F_1(D)\,u + G_1(D)\,v + H_1(D)\,w = \phi_1(x), \\ F_2(D)\,u + G_2(D)\,v + H_2(D)\,w = \phi_2(x), \\ F_3(D)\,u + G_3(D)\,v + H_3(D)\,w = \phi_3(x), \end{aligned}\right\} \tag{2.5.1}$$

in which $F_i(D)$, $G_i(D)$, $H_i(D)$ $(i = 1, 2, 3)$ are known polynomial operators in $D\ (\equiv d/dx)$, of the type discussed in section 2.1.

Because of 'linearity' of the equations, it is straight-forward to verify that the general solution of the system (2.5.1) can be expressed as:

$$(u, v, w)_{GS} = (u, v, w)_I + (u, v, w)_{II} + (u, v, w)_{III}. \tag{2.5.2}$$

where the three special solutions $(n, v, w)_{I,\,II,\,III}$ represent the general solutions of the system of equation (2.5.1) in the following three cases:

Case I : $\phi_1 \neq 0, \phi_2 = \phi_3 = 0$; *Case II* : $\phi_2 \neq 0, \phi_1 = \phi_3 = 0$ and *Case III* : $\phi_3 \neq 0, \phi_1 = \phi_2 = 0$.

It is therefore sufficient for our purpose to describe the method of determining $(u, v, w)_I$ only, since the method for the determination of the other two solutions is similar.

We consider the Determinant of the coefficients

$$\Delta(D) = \begin{vmatrix} F_1(D), & G_1(D), & H_1(D) \\ F_2(D), & G_2(D), & H_2(D) \\ F_3(D), & G_3(D), & H_3(D) \end{vmatrix} \tag{2.5.3}$$

and use small letters $f_1(D)$, $g_1(D)$ etc. to denote the co-factors of $F_1(D)$, $G_1(D)$ etc., in the expansion of $\Delta(D)$.

From the properties of determinants, we deduce that

$$f_1(D)\,F_2(D) + g_1(D)\,G_2(D) + h_1(D)\,H_2(D)$$
$$= f_1(D)\,F_3(D) + g_1(D)\,G_3(D) + h_1(D)\,H_3(D) = 0 \qquad (2.5.4)$$

and that

$$\Delta(D) = f_1(D)\,F_1(D) + g_1(D)\,G_1(D) + h_1(D)\,H_1(D). \qquad (2.5.5)$$

we thus observe that [using (2.5.4) and (2.5.5)],the solution $(u, v, w)_I$, of the system of equations (2.5.1) in the special *Case I*, i.e., when $\phi_2 = \phi_3 = 0$ and $\phi_1 \neq 0$, can be expressed in the following form:

$$u_I = f_1(D)\,V_I,\; v_1 = g_1(D)\,V_I,\; w_I = h_1(D)\,V_I, \qquad (2.5.6)$$

where V_I satisfies [substituting (2.5.6) in (2.5.1)], the single equation

$$\Delta(D)\,V_I = \phi_1(x) \qquad (2.5.6a)$$

which can be handled for solution by the methods described in *section* 2.1.

Example 2.5.1 : Solve the system of equations

$$(D+1)\,u - v = e^{-2x}$$
$$Dv + (D+1)\,w = 0,$$
$$-2u + (D-3)\,w = 0.$$

Solution

The system is already in the form of *Case 1*, described above. We get

$$\Delta(D)\,V_I \equiv (D^2-1)(D-2)\,V_I = e^{-2x}, \qquad (2.5.7)$$

with the general solution

$$V_I = Ae^{-x} + Be^{x} + Ce^{2x} - \frac{1}{12}e^{-2x}, \qquad (2.5.8)$$

where A, B, C are arbitrary constants.

Then, the solution $(u, v, w)_{GS} \equiv (u, v, w)_I$ is obtained in the following forms :

$$\left.\begin{aligned} u_{GS} &= (D^2-3D)\,V_I = 4Ae^{-x} - 2Be^{x} - 2Ce^{2x} - \frac{5}{6}e^{-2x}, \\ v_{GS} &= -(2D+2)\,V_I = -4Be^{x} - 6Ce^{2x} - \frac{1}{6}e^{-2x}, \\ w_{GS} &= 2DV_I = -2Ae^{-x} + 2Be^{x} + 4Ce^{2x} + \frac{1}{3}e^{-2x}. \end{aligned}\right\} \qquad (2.5.9)$$

Note that there are three arbitrary constants A, B, C, arising in the general solution of the above system of equations, which can be identified to be equivalent to a third order differential equation.

Example 2.5.2 : Solve the system

$$(D+1)\,u - v = e^{-2x}$$
$$Dv + (D+1)\,w = 4x$$
$$-2u + (D-3)\,w = 0.$$

Solution

To obtain the general solution of this system we have just to add a particular solution $(u, v, w)_{II}$ to the solution $(u, v, w)_I$, obtained in the previous example where

$$u_{II} = (D-3)\,V_{II},\quad v_{II} = (D+1)\,(D-3)\,V_{II},\quad w_{II} = 2V_{II}, \tag{2.5.10}$$

with

$$(D^2-1)\,(D-2)\,V_{II} = 4x. \tag{2.5.11}$$

We get the particular solution of this equation as given by

$$V_{II} = 2x + 1$$

giving

$$u_{II} = -\,6x - 1,\; v_{II} = -\,6x - 7,\; w_{II} = 4x + 2. \tag{2.5.12}$$

The more usual method of solution of systems of ordinary differential equations is by elimination, which can be effectively used only in some simple cases, an example of which is the following:

Example 2.5.3 :

$$(D+5)\,u + 2v = x$$
$$(D-1)\,u + Dv = 1.$$

Solution

Differentiating the first equation w.r.t. x and eliminating v, by using the second equation, we get

$$(D^2+3\,D+2)\,u = -\,1, \tag{2.5.13}$$

with the solution

$$u = Ae^{-x} + Be^{-2x} - \frac{1}{2}. \tag{2.5.14}$$

Then the first equation directly gives that

$$v = \frac{1}{4}\,(2x+5) - 2Ae^{-x} - \frac{3}{2}\,Be^{-2x} \tag{2.5.15}$$

and the solution of the given system of differential equations is completed.

2.5.2 The Laplace Transform Method

The Laplace-transform method is the one that is applicable to linear ordinary differential equations with constant coefficients (or to system of such equations), when the values of the unknown function $y\,(x)$ and its derivatives are assumed to be known at $x = 0$, and when we required to determine $y\,(x)$ for all $x > 0$.

We have the following basic definitions and results on Laplace Transforms (see Sneddon [13] for more details), that will be used in our discussion.

Definition 2.5.1. The 'Laplace Transform' $\overline{y}\,(p)$ of a function $y(x)$ is defined as

$$\overline{y}\,(p) = \int_0^\infty y\,(x)\,e^{-px}\,dx. \tag{2.5.16}$$

whenever the integral on the right exists.

Theorem 2.5.1

If $\bar{y}(p)$ is the Laplace transform of $y(x)$, then $\bar{y}(p+a)$ is the Laplace transform of ye^{-ax}, whenever both exist simultaneously.

Proof

$$\bar{y}(p+a) = \int_0^\infty y(x)\, e^{-(p+a)x}\, dx = \int_0^\infty (ye^{-ax})\, e^{-px}\, dx.$$

Hence the result follows:

Theorem 2.5.2

If $y, Dy, \ldots, D^{n-1}y$ are continuous functions of x, and $y_0, y_1, \ldots y_{n-1}$ are their values when $x = 0$, then

$$\overline{\left(\frac{d^r y}{dx^r}\right)}(p) \equiv [\text{Laplace Transform of } D^r y]$$

$$= p^r \bar{y}(p) - p^{r-1} y_0 - p^{r-2} y_1 - \ldots - y_{r-1}, (r = 1, 2, \ldots n). \quad (2.5.17)$$

Proof

The result follows easily by integration by parts. We only show here how to proceed in the case when $r = 2$. The general r-case can be done similarly.

$$\overline{\left(\frac{d^2 y}{dx^2}\right)}(p) = \int_0^\infty \frac{d^2y}{dx^2} e^{-px}\, dx = \left[e^{-px}\frac{dy}{dx}\right]_0^\infty + p\int_0^\infty e^{-px}\frac{dy}{dx}\, dx$$

$$= -y_1 + p\left\{[e^{-px}y]_0^\infty + p\int_0^\infty ye^{-px}\, dx\right\}$$

$$= -y_1 - py_0 + p^2\bar{y}(p).$$

Before proceeding further, we give below a short table of Laplace transforms, the entries of which can be easily verified.

Table

$y(x)$	$\bar{y}(p)$,
1	$1/p, p > 0$
x^n	$n!/p^{n+1}, n = 0, 1, 2 \ldots, p > 0$
$e^{\alpha x}$	$1/(p - \alpha), p > \alpha$
$\sin wx$	$w/(p^2 + w^2), w > 0, p > 0$
$\cos wx$	$p/(p^2 + w^2), w > 0, p > 0$
$\sinh \alpha x$	$\alpha/(p^2 - \alpha^2), \alpha > 0, p > \alpha$
$\cosh \alpha x$	$p/(p^2 - \alpha^2), \alpha > 0, p > \alpha$
$\frac{x}{2w}\sin wx$	$p/(p^2 + w^2), w > 0, p > 0$

For bigger tables of Laplace transforms, see [13].

In order to use the Laplace transform method to solve the O.D.E.

$$(D^n + a_1 D^{n-1} + \dots + a_n)\, y = \phi(x), \tag{2.5.18}$$

we take Laplace transform of both sides of (2.5.18), assuming them to exist, and obtain, by the help of Theorem 2.5.2, the transformed equation

$$(p^n + a_1 p^{n-1} + \dots + a_n)\, \bar{y}(p) = \bar{\phi}(p) + a_{n-1} y_0 + a_{n-2}(py_0 + y_1) + \dots + (p^{n-1} y_0 + p^{n-2} y_1 + \dots + y_{n-1}); \tag{2.5.19}$$

where y_0, y_1, y_{n-1} are the values of $y, Dy, \dots, D^{n-1} y$ at $x = 0$.

Equation (2.5.19) gives the Laplace transform $\bar{y}(p)$ of the actual unknown function $y(x)$ and the recovery of $y(x)$ from $\bar{y}(p)$ can be successfully completed by using the Laplace inversion formula (see [13]) or a table of Laplace transforms. [The details of Laplace inversion procedure is beyond the scope of this book and we will be satisfied for the time being with the use of a table of Laplace transforms only.] The following examples illustrate the procedure.

Example 2.5.4 : Solve $(D^2 + n^2)\, y = \sin wx\ (w \neq n)$, with $y = y_0, Dy = y_1$, when $x = 0$.

Solution

The Laplace-transformed equation (2.5.19), in this example is given by

$$(p^2 + n^2)\, \bar{y}(p) = \frac{w}{p^2 + w^2} + py_0 + y_1,$$

and this result can be expressed as

$$\bar{y}(p) = \frac{w}{w^2 - n^2}\left\{\frac{1}{p^2 + n^2} - \frac{1}{p^2 + w^2}\right\} + \frac{py_0 + y_1}{p^2 + n^2}. \tag{2.5.20}$$

To obtain $y(x)$, from (2.5.20), we use the above table and find that

$$y(x) = \frac{1}{n(w^2 - n^2)}\left[w \,.\, L^{-1}\left(\frac{n}{p^2 + n^2}\right) - nL^{-1}\left(\frac{w}{p^2 + w^2}\right)\right] + y_0 \,.\, L^{-1}\left(\frac{p}{p^2 + n^2}\right) + y_1 \,.\, L^{-1}\left(\frac{1}{p^2 + n^2}\right)$$

$$= \frac{1}{n(w^2 - n^2)}\{w \sin nx - n \sin wx\} + y_0 \cos nx + \frac{y_1}{n} \sin nx. \tag{2.5.21}$$

[Writing $L^{-1}(f(p))$ for the Laplace inversion of $f(p)$].

Example 2.5.5 : Solve the system

$$\left.\begin{aligned}(D + 1)\, y + Dz &= 0\\ (D - 1)\, y + 2Dz &= e^{-x}\end{aligned}\right\},$$

with $y = y_0$, $z = 0$, when $x = 0$.

Solution

The Laplace-transformed system is obtained as

$$\left.\begin{aligned} (p+1)\,\bar{y}(p) + p\,\bar{z}(p) &= y_0 \\ (p-1)\,\bar{y}(p) + 2p\,\bar{z}(p) &= \frac{1}{p+1} + y_0 \end{aligned}\right\} \qquad (2.5.22)$$

We observe that the given system of linear ordinary differential equations transform into a system of linear algebraic equations for the Laplace-transformed unknown functions.

Solving (2.5.22), we find that

$$\bar{y}(p) = \frac{y_0}{p+3} - \frac{1}{(p+1)(p+3)} \qquad (2.5.23)$$

and

$$\bar{z}(p) = \frac{y_0}{p} - y_0\frac{p+1}{(p+3)\,p} + \frac{1}{(p+3)\,p}. \qquad (2.5.24)$$

The actual unknown $y(x)$ and $z(x)$ can be recovered by first expressing the right sides of the equations (2.5.23) and (2.5.24) in convenient forms, by using partial fractions rules of algebra, and then using the table of Laplace transforms.

We find that

$$y(x) = y_0 L^{-1}\left(\frac{1}{p+3}\right) - \frac{1}{2}L^{-1}\left(\frac{1}{p+1}\right) + \frac{1}{2}L^{-1}\left(\frac{1}{p+3}\right)$$

i.e., $$y(x) = \left(y_0 + \frac{1}{2}\right)e^{-3x} - \frac{1}{2}e^{-x}, \qquad (2.5.25)$$

and

$$z(x) = \left(\frac{2y_0+1}{3}\right)L^{-1}\left(\frac{1}{p} - \frac{1}{p+3}\right)$$

i.e., $$z(x) = \left(\frac{2y_0+1}{3}\right)(1 - e^{-3x}). \qquad (2.5.26)$$

It can be easily checked, to gain confidence, that (2.5.25) and (2.5.26) really solve the given system of ordinary differential equations.

PROBLEMS

1. Find the general solutions of the following equations ($D = d/dx$):
   (i) $(D^3 + D^2 - 2)\,y = x^2 - 1$
   (ii) $(4D^4 - 20\,D^2 + 25)\,y = x + e^{2x}$

(iii) $(D^2 + 1)\, y = \sin x \sin 2x$

(iv) $(D^2 - D + 1)\, y = \sin^2 x \cos^2 x$

(v) $(D^3 - 8)\, y = \cosh x$

(vi) $(D^3 - 3D + 2)\, y = \cosh^2 x$

(vii) $(D - 1)^2\, y = x^2 \sinh 2x$

(viii) $(x^4 D^4 + 5x^3 D^3 + x^2 D^2 + 2xD - 2)\, y = x^2$

(ix) $x^2 y'' - 7xy' + 15y = 0 \left(' = \dfrac{d}{dx} \right)$

2. Verify that e^{x^2} is a solution of the equation

$$y^{(2)} - 4xy^{(1)} + (4x^2 - 2)\, y = 0$$

and hence obtain its general solution.

3. One solution of the equation $x^2 y^{(2)} - xy^{(1)} + y = 0$ is $y_1(x) = x$. Find the solution of

$$x^2 y^{(2)} - xy^{(1)} + y = x^2,$$

satisfying the conditions $y(1) = 1$, $y^{(1)}(1) = 0$.

4. Find the general solution of the equation

$$x^2 y^{(2)} + 4xy^{(1)} + (2 + x^2)\, y = x^2, \text{ for } x > 0.$$

5. Let 'b' be a continuous function on an interval I, and let x_0 be a fixed point in I. Show that the function ϕ defined by

$$\phi(x) = e^{ax} \int_{x_0}^{x} \frac{(x - t)^{k-1}}{(k-1)!} e^{-at}\, b(t)\, dt,$$

'k' being a positive integer, and 'a' being a constant, satisfies the relation

$$(D - a)^k (\phi) = b.$$

6. The equation $y^{(1)} + a(x)\, y = 0$ has for a solution

$$\phi(x) = \exp\left[- \int_{x_0}^{x} a(t)\, dt \right],$$

(Here 'a' is continuous on an interval containing the point x_0). This suggests trying to find a solution of the equation

$$y^{(2)} + a_1(x)\, y^{(1)} + a_2(x)\, y = 0$$

of the form

$$\phi(x) = \exp\left[\int_{x_0}^{x} p(t)\, dt \right],$$

where p is a function to be determined. Show that ϕ is a solution of the last equation if, and only if, p satisfies the following first-order *non-linear* equation (known as *Riccati Equation*):

$$p^{(2)} = -p^2 - a_1(x)\,p - a_2(x).$$

7. Two solutions of the equation

$$x^3 y^{(3)} - 3xy^{(1)} + 3y = 0 \ (x > 0)$$

are $y_1(x) = x^2$ and $y_2(x) = x^3$. Use this information to find a third independent solution.

8. Solve the following system of equations:

(i) $(D-1)\,u - 2v = e^{-x}$; $-2u + (D-1)\,v = 1$.

(ii) $Du - v = 0$; $Dv - w = 0$, $Dw - u = 0$.

(iii) $(D^2 - 4D)\,u - (D-1)\,v = e^{4x}$;

$(D+6)\,u + (D^2 - D)\,v = 0$.

9. Solve the following with initial values $y_0, y_1 \ldots$ of $y, Dy, \ldots$ etc.

(i) $(D+1)(D+2)\,y = e^{-x}$,

(ii) $(D^2 + n^2)\,y = \sin nx$,

(iii) $(D+1)\,u - (5D+7)\,v = 1$; $u + (D-1)\,v = 0$

10. (i) Show that if $\bar{y}(p)$ is the Laplace transform of $y(x)$, $\bar{y}(p)/p$ is the Laplace transform of

$$\int_0^x y(\xi)\,d\xi.$$

(ii) Show that if $\bar{y}(p)$ and $\bar{z}(p)$ are the Laplace transforms of $y(x)$ and $z(x)$, then $\bar{y}(p)\,\bar{z}(p)$ is the Laplace transform of

$$\int_0^x y(x-\xi)\,z(\xi)\,d\xi = \int_0^x y(\xi)\,z(x-\xi)\,d\xi.$$

[This result is known as the 'Convolution Theorem' and is extremely useful in application.]

CHAPTER - THREE

Series Solution Near An Ordinary Point: Hermite Functions

In this Chapter we shall describe a method of obtaining solutions of linear ordinary differential equations in the form of an infinite series, that is valid in the 'neighbourhood' of an 'Ordinary Point' of the given differential equation. The utility of this method is then shown in the case of a special equation, known as Hermite's equation, and the first 'special function' to be described in this book is defined and certain useful properties of this function are derived.

3.1 SERIES SOLUTION NEAR AN ORDINARY POINT

Without any loss of generality, we shall consider here, the following differential equation:

$$y^{(n)} + p_1(x)\, y^{(n-1)} + p_2(x)\, y^{(n-2)} + \dots + p_n(x)\, y = Q(x) \qquad (3.1)$$

and assume that the coefficients $p_i(x)$ $(i = 1, 2, \dots, n)$ as well as the right hand side $Q(x)$ are 'analytic' functions of the real variable x in a neighbourhood of any fixed point x_0, so that the point $x = x_0$ is an Ordinary Point of equation (3.1).

Definition 3.1

A function $f(x)$ is said to be analytic in a neighbourhood of a point x_0, if derivatives of all orders exist, for the function $f(x)$, in that neighbourhood and if $f(x)$ also prossesses a convergent Taylor series there e.g., x^n, e^x, $\sin x$, $\cos x$, with n as a positive integer, are analytic functions of x in some neighbourbood of any given point x_0.

The following theorem can be proved easily.

Theorem 3.1

The solution of the differential equation (3.1) is analytic in the neighbourhood of the point x_0, in which the coefficients $p_i(x)$ of the equation are known to be analytic.

Proof

Assuming that $y = f(x)$ is a possible solution of equation (3.1), we write

$$y^{(j)}(x_0) = f^{(j)}(x_0) = f_{0,j}, j = 0, 1, 2, \ldots, n-1. \tag{3.2}$$

Then, substituting from (3.2) in the given equation (3.1), we can determine the values of $y^{(n)}(x_0)$, $y^{(n+1)}(x_0)$, etc. by repeated differentiations. We thus observe that derivatives of all orders exist finitely for any solution $y = f(x)$, of the given equation (3.1) once the coefficients $p_i(x)$ are assumed to be analytic in a neighbourhood of the point x_0.

It, therefore, only remains to prove that the Taylor series for the solution $y = f(x)$ converges in the same neigbourhood of x_0, in which the coefficient functions $p_i(x)$ are also analytic.

To this end we assume that

$$p_i(x) = \sum_{j=0}^{\infty} \frac{p_i^{(j)}(x_0)}{j!}(x - x_0)^j, \tag{3.3}$$

so that we get,

$$p_i^{(k)}(x) = \sum_{j=k}^{\infty} \frac{p_i^{(j)}(x_0)}{(j-k)!}(x - x_0)^{j-k}. \tag{3.4}$$

Now, equation (3.1) gives that

$$y^{(n)} = Q - \sum_{i=1}^{n} p_i(x)\, y^{(n-i)}. \tag{3.5}$$

Therefore, we obtain

$$y^{(n+k)} = Q^{(k)} - \sum_{i=1}^{n} p_i^{(k)}(x)\, y^{(n-i)} - \sum_{i=1}^{n} p_i(x)\, y^{(n+k-i)} \tag{3.6}$$

so that

$$y^{(n)}(x_0) \equiv f^{(n)}(x_0) = -\sum_{i=1}^{n} p_i(x_0) f_{0,n-i} + Q(x_0) \tag{3.7}$$

and

$$y^{n+k}(x_0) = f^{(n+k)}(x_0)$$

$$= Q^{(k)}(x_0) - \sum_{i=1}^{n} p_i(x_0) f_{0,n+k-i} - \sum_{i=1}^{n} p_i^{(k)}(x_0) f_{0,n-i}. \quad (k \geq 1) \tag{3.8}$$

Thus the Taylor series for $y(x)$ takes up the form

$$\sum_{j=0}^{\infty} y^{(j)}(x_0)\frac{(x-x_0)^j}{j!}$$

$$= \sum_{j=0}^{n-1} f_{0,j}\frac{(x-x_0)^j}{j!} - \frac{(x-x_0)^n}{n!}\left(\sum_{i=1}^{n} p_i(x_0) f_{0,n-i} - Q(x_0)\right)$$

$$+ \sum_{k=1}^{\infty} Q^{(k)}(x_0)\frac{(x-x_0)^{n+k}}{(n+k)!} - \sum_{k=1}^{\infty}\frac{(x-x_0)^{n+k}}{(n+k)!}$$

$$\times\left\{\sum_{i=1}^{n} p_i(x_0) f_{0,n+k-i} + \sum_{t=1}^{n} p_i^{(k)}(x_0) f_{0,n-i}\right\} \tag{3.9}$$

and its convergence in a neighbourhood $| x - x_0 | < h$, for a given finite h, is apparent now if a rearrangement of the last two series are carried out along with the observation that all the derivatives $f_{0,p}$ $(p \geq 0)$ are finite.

The above theorem serves as a very useful tool to solve the Ordinary Differential equation (3.1), in a neighbourhood of an ordinary point x_0, in the form of a convergent series.

The method is as follows:

Assume the general solution of equation (3.1) in the form:

$$y(x) = \sum_{s=0}^{\infty} c_s (x-x_0)^s, \tag{3.10}$$

substitute in the given equation (3.1) and equate the coefficients of identical powers of $x - x_0$ from both sides of the resulting algebraic equation to get a set of relations (recurrence relations) connecting the unknown coefficients c_s. It then so happens that all the coefficients, after the first n (i.e. $c_0, c_1, \ldots, c_{n-1}$), can be determined in terms of these first n coefficients. Finally, substituting the values of c_s so obtained in the form (3.10) of the solution of equation (3.1) we find that the general solution can be determined involving 'n' arbitrary constants $c_0, c_1, \ldots, c_{n-1}$.

We consider the following example which illustrates the methods very clearly.

Example 3.1.1 : Solve, in the neighbourhood of $x = 0$:

$$\frac{d^2 y}{dx^2} + 2x\frac{du}{dx} - 2y = 1$$

Solution

We first observe that $x = 0$ is an ordinary point of the given differential equation and then, that the coefficients $2x$, -2 and the right hand side, 1, are all analytic functions of x in any finite neighbourhood of the point $x = 0$.

Thus, the theory described above, is applicable to the problem and we assume the general solution in the form

$$y(x) = \sum_{s=0}^{\infty} c_s x^s.$$

Substituting in the given equation we get

$$\sum_{s=0}^{\infty} c_s [s(s-1)x^{s-2} + 2sx^s - 2x^s] = 1$$

or

$$\sum_{s=0}^{\infty} c_s [s(s-1) + 2(s-1)x^2] x^s = x^2,$$

or

$$\sum_{s=2}^{\infty} c_s s(s-1) x^s + 2 \sum_{s=0}^{\infty} c_s (s-1) x^{s+2} = x^2, \qquad (*)$$

(Assuming c_0, c_1 to be arbitrary).

Equating the coefficients of identical powers of x from both sides of equation (*) we get,

$$x^2 : c_2 . 2.1 + 2 . c_0 (-1) = 1; \; c_2 = c_0 + \frac{1}{2}$$

$$x^3 : c_3 . 3.2 + 2 c_1 . 0 = 0; \; c_3 = 0$$

$$x^n : c_n . n(n-1) + 2 c_{n-2} (n-3) = 0;$$

giving the "recurrence relation"

$$c_n = -2 \frac{(n-3)}{n(n-1)} c_{n-2}; \quad \text{for } n > 3$$

so that, we obtain,

$$c_3 = c_5 = c_7 = \dots = 0, \text{(odd coefficients)}$$

$$c_2 = c_0 + \frac{1}{2},$$

$$c_4 = -2 . \frac{1}{4.3} c_2 = -\frac{1}{6} c_2 = -\frac{1}{12} - \frac{c_0}{6},$$

$$c_6 = -2 . \frac{3}{6.5} c_4 = -\frac{c_4}{5} = \frac{1}{60} + \frac{c_0}{30},$$

etc.

Thus we obtain the general solution of the given second order differential equation in the form:

$$y = c_0 \left[1 + x^2 - \frac{x^4}{6} + \frac{x^6}{30} - \dots \right] + c_1 x + \left[\frac{x^2}{2} - \frac{x^4}{12} + \frac{x^6}{60} - \dots \right], \qquad (**)$$

involving only two arbitrary constants c_0 and c_1.

Note: It is interesting to verify that $y = x$ is a solution of the homogeneous part of the given equation and this fact is also clear from the form (**) of the general solution of the equation.

3.2 HERMITE'S DIFFERENTIAL EQUATION

Definition 3.2.1.

The linear Ordinary Differential equation

$$\frac{d^2 y}{dx^2} - 2x\frac{dy}{dx} + 2\nu y = 0 \tag{3.11}$$

with ν as a known constant (real or complex), is called 'Hermite's Equation'.

It is immediately observed that $x = 0$ is an 'Ordinary Point' of Hermite's equation and hence, its general solution can be determined by using the method described in the previous section.

Assuming the solution of equation (3.11) in the form

$$y(x) = \sum_{s=0}^{\infty} c_s x^s, \tag{3.12}$$

we obtain

$$\sum_{s=0}^{\infty} [c_{s+2}(s+2)(s+1) - 2c_{s+1}(s+1)x + 2\nu c_s]x^s = 0 \tag{3.13}$$

from which, by equating the coefficients of all powers of x to zero, we arrive at the following relations connecting the constants c_s:

$$\left.\begin{array}{ll} x^0: 2c_2 + 2\nu c_0 = 0; & c_2 = -\nu c_0 = -\dfrac{2\nu}{2!}c_0 \\ x^1: 3!\,c_3 - 2c_1 + 2\nu c_1 = 0; & c_3 = -\dfrac{2(\nu-1)}{3!}c_1 \\ x^2: c_4.(4.3) - 2c_2.2 + 2\nu c_2 = 0; & c_4 = \dfrac{2c_2(2-\nu)}{4.3}, \\ \quad c_4 = +\dfrac{2^2.\nu(\nu-2)}{4!}c_0 & \\ x^3: c_5.5.4 - 2c_3.3 + 2\nu c_3 = 0; & c_5 = \dfrac{2c_3(3-\nu)}{5.4} = \dfrac{2^2(\nu-1)(\nu-3)}{5!}c_1 \end{array}\right\}$$

etc. (3.14)

Substituting from the relations (3.14), the values of the constants $c_2, c_3, c_4\, c_5, \ldots$, in terms of *first two* constants c_0 and c_1, into equation (3.12), we obtain the general solution of Hermite's equation in the form:

$$y(x) = c_0 y_1(x) + c_1 y_2(x), \tag{3.15}$$

where

$$y_1(x) = 1 - \frac{2\nu}{2!}x^2 + \frac{2^2 \nu(\nu-2)}{4!}x^4 - \frac{2^3 . \nu(\nu-2)(\nu-4)}{6!}x^6 + \ldots$$

and

$$y_2(x) = x - \frac{2(\nu-1)}{3!}x^3 + \frac{2^2(\nu-1)(\nu-3)}{5!}x^5 + \ldots \tag{3.16}$$

3.2.1 Hermite's Polynomial

It is easily observed that the two functions $y_1(x)$ and $y_2(x)$, as given above, are two linearly independent solutions of Hermite's equation.

It is also observed that in the case when the constant ν represents a positive integer 'n', then one of the two solutions y_1 or y_2 reduces to a polynomial of degree 'n', according as 'n' is even or odd.

By making suitable choices of the constants c_0 and c_1 we are, then, motivated to define the following important polynomial, known as 'Hermite's polynomial'.

Definition 3.2.2

The Hermite's polynomial $H_n(x)$ of degree 'n', is defined by the formula:

$$H_n(x) = 2^n x^n - \frac{2^{n-2}}{1!}n(n-1)x^{n-2} + \frac{2^{n-4}}{2!}n(n-1)(n-2)(n-3)x^{n-4}$$
$$- \frac{2^{n-6}}{3!}n(n-1)(n-2)(n-3)(n-4)(n-5)x^{n-6} + \ldots \tag{3.17}$$

The first few Hermite's polynomials are given by

$$H_0(x) = 1, \; H_1(x) = 2x,$$
$$H_2(x) = 4x^2 - 2, \; H_3(x) = 8x^3 - 12x, \text{ etc.} \tag{3.18}$$

3.2.1 (a) Generating function for Hermite's polynomial

The following important result can be easily derived

Theorem 3.2.1

For all real values of the variables x and t, the function

$$f(x, t) = \exp[(2tx - t^2)],$$

has the following expansion in terms of Hermite's polynomials $H_n(x)$:

$$f(x, t) = \sum_{n=0}^{\infty} \frac{t^n}{n!} H_n(x). \tag{3.19}$$

Proof

Expanding the functions e^{2tx} and e^{-t^2} in a usual manner, we get,

$$f(x, t) = e^{2tx} . e^{-t^2}$$

$$= \left[1 + 2tx + \frac{2^2 (tx)^2}{2!} + \frac{2^3 (tx)^3}{3!} + \ldots + \frac{2^n (tx)^n}{n!} + \ldots\right]$$

$$\times \left[1 - t^2 + \frac{t^4}{2!} - \frac{t^6}{3!} + \ldots + \frac{(-1)^n t^{2n}}{n!} + \ldots\right]. \tag{3.20}$$

Collecting all the coefficients of t^n, for a fixed n, from the product on the right hand side of equation (3.20) we find that

$$f(x, t) = \sum_{n=0}^{\infty} t^n H_n(x)/n!,$$

and the theorem is proved.

Definition 3.2.3

The function $\exp(2tx - t^2)$ is called the 'generating function' for Hermite's polynomials.

3.2.1 (b) Rodrigue's formula

Theorem 3.2.2

Hermite's polynomial $H_n(x)$ satisfies the following formula, known as *Rodrigue's formula for Hermite's polynomials*

$$H_n(x) = (-1)^n e^{x^2} \frac{d^n}{dx^n} (e^{-x^2}). \tag{3.21}$$

Proof

Observing that the function $e^{2tx - t^2}$ is analytic in any neigbourhood of the point $t = 0$, for any fixed value of x, we have the following Taylor series representation valid:

$$\exp(2tx - t^2) = \sum_{n=0}^{\infty} \left[\frac{\partial^n}{\partial t^n} (\exp\{2tx - t^2\})\right]_{t=0} \frac{t^n}{n!}. \tag{3.22}$$

Then using the result that

$$\exp(2tx - t^2) = e^{x^2} . \exp(-(x - t)^2)$$

we find that

$$\frac{\partial^n}{\partial t^n} (\exp\{2tx - t^2\}) = e^{x^2} \frac{\partial^n}{\partial t^n} (\exp\{-(x - t)^2\}). \tag{3.23}$$

Now, we have that

$$\frac{\partial}{\partial t} (\exp\{-(x - t)^2\}) = 2(x - t) \exp(-(x - t)^2)$$

and

$$\frac{\partial}{\partial x}(\exp\{-(x-t)^2\}) = -2(x-t)\exp(-(x-t)^2),$$

so that we have the identity

$$\frac{\partial}{\partial t}(\exp\{-(x-t)^2\}) = -\frac{\partial}{\partial x}(\exp\{-(x-t)^2\})$$

and the repeated use of it gives that

$$\frac{\partial^n}{\partial t^n}(\exp\{-(x-t)^2\}) = (-1)^n \frac{\partial^n}{\partial x^n}(\exp\{-(x-t)^2\}). \tag{3.24}$$

Using the results (3.23) and (3.24) in (3.22), we thus obtain

$$\exp(2tx - t^2) = e^{x^2}(-1)^n \sum_{n=0}^{\infty} \frac{d^n}{dx^n}(e^{-x^2})\frac{t^n}{n!}. \tag{3.25}$$

Comparing the two results (3.25) and (3.19), we ultimately prove the result (3.21).

3.2.1 (c) Recurrence relations

Hermite's polynomials can be shown to satisfy certain very useful 'recurrence relations' by exploiting the results obtained in the theorems 3.2.1 and 3.2.2.

Theorem 3.2.3

The following basic 'recurrence relation' for Hermite's polynomials hold good:

(i) $H'_n(x) = 2n H_{n-1}(x)$

and

(ii) $2x H_n(x) = 2n H_{n-1}(x) + H_{n+1}(x)$,

where dash denotes differentiation.

Proof

We have the result (Theorem 3.2.1):

$$\exp(2tx - t^2) = \sum_{n=0}^{\infty} H_n(x)\frac{t^n}{n!}. \tag{3.26}$$

(i) Differentiating both sides of (3.26) with respect to (w.r.t.) x, we get

$$2t\exp(2tx - t^2) = \sum_{n=0}^{\infty} H'_n(x)\frac{t^n}{n!},$$

or, using (3.26) again, we get

$$2t\sum_{n=0}^{\infty} H_n(x)\frac{t^n}{n!} = \sum_{n=0}^{\infty} H'_n(x)\frac{t^n}{n!},$$

Comparing the coefficients of t^n from both sides we thus obtain

$$2\,n\,H_{n-1}(x) = H'_n(x) \tag{3.27}$$

and this is the result (i).

(ii) Differentiating both sides of (3.26) w.r.t t, we similarly derive that

$$2(x-t)\sum_{n=0}^{\infty} H_n(x)\frac{t^n}{n!} = \sum_{n=0}^{\infty} H_{n+1}(x)\frac{t^n}{n!}.$$

Collecting the coefficients of t^n from both sides, again, we get,

$$2\,x\frac{H_n(x)}{n!} - 2\frac{H_{n-1}(x)}{(n-1)!} = \frac{H_{n+1}(x)}{n!}$$

or,

$$2\,x\,H_n(x) = 2\,n\,H_{n-1}(x) + H_{n+1}(x), \tag{3.28}$$

and this is the result (ii).

Eliminating $H_{n-1}(x)$ between the above two results (i) and (ii) we obtain

$$H_n'(x) = 2\,x\,H_n(x) - H_{n+1}(x). \tag{3.29}$$

Differentiating (3.29) w.r.t. x we get

$$H_n''(x) = 2\,H_n(x) + 2\,x\,H_n'(x) - H_{n+1}'(x)$$

$$= 2\,H_n(x) + 2\,x\,H_n'(x) - 2(n+1)\,H_n(x) \quad \text{(by using (3.27))}.$$

We thus verify that $H_n(x)$ satisfies Hermite's differential equation:

$$\frac{d^2y}{dx^2} - 2\,x\frac{dy}{dx} + 2\,ny = 0 \tag{3.30}$$

where n is a positive integer.

3.3 HERMITE'S FUNCTION

Definition 3.3.1

The function $\psi_n(x)$ defined by the relation

$$\psi_n(x) = e^{-1/2\,x^2} H_n(x), \tag{3.31}$$

where $H_n(x)$ is Hermite's polynomial of degree 'n' is called Hermite's function of order 'n'.

By utilizing the various relations, derived earlier, for Hermite's polynomials, we can easily prove the following results:

Theorem 3.3.1

The Hermite's functions $\psi_n(x)$ satisfy the relations:

(i) $2\,n\,\psi_{n-1} = x\,\psi_n + \psi_n'$,

(ii) $2x\psi_n = 2n\psi_{n-1} + \psi_{n+1}$,

(iii) $\psi_n' = x\psi_n - \psi_{n+1}$,

(iv) $\psi_n = (-1)^n e^{1/2 x^2} \dfrac{d^n}{dx^n}(e^{-x^2})$,

'dash' denoting differentiation w.r.t. x.

Proof

Using definition (3.3.1) in the recurrence relations (3.27) and (3.28) directly, we prove the recurrence relations (i) and (ii) for $\psi_n(x)$.

Eliminating ψ_{n-1} between (i) and (ii), the relation (iii) follows easily. The result (iv) can be deduced by using Rodrigue's formula (3.21) along with definition (3.3.1).

3.3.1 Orthogonality Property of Hermite's Functions

We first define the concept of orthogonal sets of functions.

Definition 3.3.2

A set of real-valued functions $\{\phi_m(x)\}_{m=1,2,\ldots}$ for the real variable $x \in (a, b)$, is said to be an *orthogonal* set in the interval (a, b) with the weight function $w(x)$, if the following relation holds goods:

$$\int_a^b \phi_m(x)\,\phi_p(x)\,w(x)\,dx = A_m\,\delta_{mp}, \tag{3.32}$$

where A_m is a constant and δ_{mp} is the Krönecker Delta ($\delta_{mp} = 0$, if $m \neq p$, and $= 1$, if $m = p$). If $w(x) = 1$, then the set $\{\phi_m\}$ is said to be 'orthogonal' in the interval (a, b).

The following theorem can be proved as follows:

Theorem 3.3.2

$$\int_{-\infty}^{\infty} \psi_m(x)\,\psi_n(x)\,dx = A_n\,\delta_{mn}, \tag{3.33}$$

where ψ_m is Hermite's function of order 'm', and

$$A_n = 2^n \,.\, n!\,\sqrt{\pi}, \tag{3.34}$$

δ_{mn} being the Krönecker delta.

[i.e., The set of Hermite's functions $\{\psi_m(x)\}_{m=1,2,\ldots}$ is an orthogonal set in the interval $(-\infty, \infty)$.]

Proof

Using first the definition (3.3.1) and the fact that $H_n(x)$ satisfies Hermite's equation (3.30), we find that $\psi_n(x)$ satisfies the ordinary differential equation

$$\psi_n'' + (2n + 1 - x^2)\,\psi_n = 0 \tag{3.35}$$

Similarly, we have that

$$\psi_m'' + (2m + 1 - x^2)\psi_m = 0. \tag{3.36}$$

Multiplying equation (3.35) by ψ_m and equation (3.36) by ψ_n, and taking difference, we get,

$$\begin{aligned} 2(m-n)\psi_m\psi_n &= \psi_n''\psi_m - \psi_m''\psi_n \\ &= \frac{d}{dx}[\psi_n'\psi_m - \psi_n\psi_m']. \end{aligned}$$

We thus have that

$$\begin{aligned} 2(m-n)\int_{-\infty}^{\infty}\psi_m(x)\psi_n(x)\,dx &= \lim_{x\to\infty}[\psi_n'\psi_m - \psi_n\psi_m'] \\ &\quad - \lim_{x\to-\infty}[\psi_n'\psi_m - \psi_n\psi_m']. \\ &= 0 - 0 = 0, \end{aligned} \tag{3.37}$$

if the definition (3.31) is utilized in calculating the limits.

It is, therefore, immediate that

$$I_{m,n} \equiv \int_{-\infty}^{\infty}\psi_m(x)\psi_n(x)\,dx = 0, \quad \text{if } m \neq n, \tag{3.38}$$

and one part of the theorem is proved. In order to prove the remaining part of the theorem we must prove that

$$I_{n,n} = 2^n \,.\, n!\sqrt{\pi}. \tag{3.39}$$

To this end, we consider the evaluation of the following integral:

$$I = \int_{-\infty}^{\infty} 2x\,\psi_n(x)\,\psi_{n-1}(x)\,dx, \tag{3.40}$$

and find, by using the recurrence relation (ii) of Theorem 3.3.1, that

$$I = 2n\,I_{n-1,\,n-1} + I_{n-1,\,n+1}$$

or

$$I = 2n\,I_{n-1,\,n-1} \tag{3.41}$$

by using the result (3.38).

Also, by integrating by parts directly, after using the result (iv) of Theorem 3.3.1, we find that

$$I = (-1)^{2n-1}\int_{-\infty}^{\infty} 2x\left(e^{1/2x^2}\frac{d^n}{dx^n}(e^{-x^2})\,e^{x^2/2}\left(\frac{d^{n-1}}{dx^{n-1}}\right)(e^{-x^2})\right)dx$$

$$= -\int_{-\infty}^{\infty} 2\, xe^{x^2} \frac{d^n}{dx^n}(e^{-x^2}) \frac{d^{n-1}}{dx^{n-1}}(e^{-x^2})\, dx$$

$$= \int_{-\infty}^{\infty} e^{x^2} \frac{d}{dx}\left[\frac{d^n}{dx^n}(e^{-x^2}) \frac{d^{n-1}}{dx^{n-1}}(e^{-x^2})\right] dx + 0, \quad \text{(integrating by parts)}$$

$$= \int_{-\infty}^{\infty} e^{x^2}\left[\frac{d^{n+1}}{dx^{n+1}}(e^{-x^2}) \frac{d^{n-1}}{dx^{n-1}}(e^{-x^2}) + \frac{d^n}{dx^n}(e^{-x^2}) \frac{d^n}{dx^n}(e^{-x^2})\right] dx$$

$$= \int_{-\infty}^{\infty} \psi_{n+1}\, \psi_{n-1}\, dx + \int_{-\infty}^{\infty} \psi_n\, \psi_n\, dx.$$

$$\therefore \qquad I = I_{n+1,\, n-1} + I_{n,\, n} = I_{n,\, n}, \tag{3.42}$$

by using the result (3.38).

Using the results (3.41) and (3.42) we thus arrive at the following recurrence relation for the integral $I_{n,\, n}$:

$$I_{n,\, n} = 2\, n\, I_{n-1,\, n-1}. \tag{3.43}$$

By repeated use of this relation, n-times, we finally obtain,

$$I_{n,\, n} = 2^n \,.\, n!\, I_{0,\, 0}$$

$$= 2^n\, n! \int_{-\infty}^{\infty} \psi_0\, \psi_0\, dx$$

$$= 2^n \,.\, n! \int_{-\infty}^{\infty} e^{-x^2}\, dx$$

i.e.,

$$I_{n,\, n} = 2^{\,n} \,.\, n!\, \sqrt{\pi} \tag{3.44}$$

Using the results (3.40), (3.42) and (3.44), the theorem is proved completely.

We end this chapter by introducing the concept of 'orthonormal' functions in the following manner:

Definition 3.3.3

If, in equation (3.32) of the definition 3.3.2, the constant A_m is replaced by 'unity', then the corresponding set $\{\phi_m\}$ is said to be an Orthonormal set with the weight $w(x)$, and if further, $w(x) = 1$, then the set is simply said to be 'Orthonormal' in the interval (a, b).

Using this definition of orthonormal sets of functions and using the results (3.33) and (3.34),we can easily prove the following theorem:

Theorem 3.3.3

The set of functions $\{\chi_n(x)\}$, where $\chi_n(x)$ is defined by the relation $\chi_n(x) = \dfrac{\psi_n(x)}{2^{n/2}\sqrt{n!}\,\pi^{1/4}}$, is an orthonormal set in the interval $(-\infty, \infty)$.

Definition 3.3.4

The function $\chi_n(x)$ is called 'normalised Hermite function'.

Of many uses of 'Orthogonal' and 'Orthonormal' sets of functions $\{\phi_m(x)\}$, the following result on generalized Fourier series is highly important and interesting in the context of Hermite-function-theory, which we shall quote here, without proof (see Ritt [11]).

Theorem 3.3.4

If a function $f(x)$ is such that it is continuous in the interval $(-\infty, \infty)$ and if $\int_{-\infty}^{\infty} |f|^2\, dx < \infty$, then $f(x)$ can be expanded in a convergent Fourier series of the form:

$$f(x) = \sum_{n=0}^{\infty} c_n \chi_n(x) \tag{3.45}$$

where the Fourier coefficients c_n are given by

$$c_n = \int_{-\infty}^{\infty} f(x)\, \chi_n(x)\, dx \tag{3.46}$$

As an application of the orthonormal Hermite's functions we consider the following example:

Example 3.3.1

If $$K(x, y, t) = \sum_{n=0}^{\infty} \chi_n(x)\, \chi_n(y)\, t^n,$$

then show that

$$\int_{-\infty}^{\infty} \exp\left(-\frac{1}{2}x^2 + 2\lambda x - \lambda^2\right) K(x, y, t)\, dx$$

$$= \exp\left(-\frac{1}{2}y^2 + 2\lambda ty - \lambda^2 t^2\right)$$

where λ is real.

Solution

Using the generating function representation for Hermite's polynomials, as given by Theorem 3.2.1 and the definition $\chi_n(x)$ as given in *definition* 3.3.3, we easily obtain

$$\exp\left(-\frac{1}{2}x^2 + 2\lambda x - \lambda^2\right) = \sum_{n=0}^{\infty} \frac{2^{1/2\, n}}{(n!)^{1/2}} \pi^{1/4} \,.\, \lambda^n \chi_n(x)$$

$$= \sum_{n=0}^{\infty} c_n \lambda^n \chi_n(x) \quad \text{(say)}. \tag{*}$$

Then, by using the Fourier series relations (3.45) and (3.46) we get

$$c_n \lambda^n = \int_{-\infty}^{\infty} \exp\left(-\frac{1}{2}x^2 + 2\lambda x - \lambda^2\right)\chi_n(x)\,dx. \qquad (**)$$

Now the given integral

$$\int_{-\infty}^{\infty} \exp\left(-\frac{1}{2}x^2 + 2\lambda x - \lambda^2\right)K(x, y, t)\,dx$$

$$= \sum_{n=0}^{\infty} \chi_n(y)\, t^n \int_{-\infty}^{\infty} \exp\left(-\frac{1}{2}x^2 + 2\lambda x - \lambda^2\right)\chi_n(x)\,dx$$

$$= \sum_{n=0}^{\infty} c_n (\lambda t)^n \chi_n(y), \text{ by using } (**)$$

$$\equiv \exp\left(-\frac{1}{2}y^2 + 2\lambda ty - \lambda^2 t^2\right), \text{ by using } (*).$$

Hence follows the result.

PROBLEMS

1. Find the general solutions of the following differential equations, in the neighbourhood of the points mentioned in the brackets, next to the equations.

   (i) $\dfrac{d^2y}{dx^2} + 3x^2\dfrac{dy}{dx} - xy = 0;\ (x = 0)$

   (ii) $\dfrac{d^2y}{dx^2} + (x-1)^2\dfrac{dy}{dx} - (x-1)y = 0;\ (x = 1)$

   (iii) $\dfrac{d^2y}{dx^2} + 2x\dfrac{dy}{dx} + e^x y = x^2 + 2x;\ (x = 0)$

   (iv) $x\dfrac{d^2y}{dx^2} - \dfrac{dy}{dx} + 2\log x \,.\, y = x^2 + 2x;\ (x = 1)$

2. Compute the solution $y = \phi(x)$ of the equation

$$\frac{d^3y}{dx^3} - xy = 0,$$

   which satisfies $\phi(0) = 1$, $\phi'(0) = 0$, $\phi''(0) = 0$ 'dashes' denoting differentiations w.r.t. x.

3. Find a series which satisfies the differential equation

$$(1 + x)\frac{dy}{dx} = my, \ (m = a \text{ constant}).$$

Prove that if $f(m)$ is the solution of the differential equation which reduces to unity when $x = 0$ then, for all values of x,

$$f(m_1)\,f(m_2) = f(m_1 + m_2).$$

4. Prove that, if $m < n$.

$$\frac{d^m}{dx^m}\{H_n(x)\} = \frac{2^m \,.\, n!}{(n-m)!} H_{n-m}(x).$$

5. Prove that

$$\int_{-\infty}^{\infty} x\,\psi_m(x)\,\psi_n(x)\,dx = \begin{cases} 0, & \text{if } m \neq n-1 \text{ or } n+1 \\ 2^n\,(n+1)\,!\,\pi^{1/2}, & \text{if } m = n+1 \\ 2^{n-1}\,(n!)\,\pi^{1/2}, & \text{if } m = n-1. \end{cases}$$

6. Prove that

$$\int_{-\infty}^{\infty} \psi_m(x)\,\psi'_n(x)\,dx = \begin{cases} 0, & \text{if } m \neq n-1 \text{ or } n+1 \\ 2^{n-1}\,(n!)\,\pi^{1/2}, & \text{if } m = n-1 \\ -2^n\,.\,(n+1)\,!\,\pi^{1/2}, & \text{if } m = n+1. \end{cases}$$

7. Show that

$$\int_{-\infty}^{\infty} x^2\,e^{-x^2}\,\{H_n(x)\}^2\,dx = (\pi^{1/2})\,.\,2^n\,(n!)\,.\,(n+1/2).$$

CHAPTER - FOUR

Series Solution Near a Regular Singular Point

In the *definition* 1.2.2 of Chapter I, we have introduced the concept of 'singular points' of a linear ordinary differential equation. In the present chapter we shall concentrate upon a special type of a singular point, known as 'Regular Singular Point', and describe the method of solution of ordinary differential equations valid near regular singular points. We start with the following basic definitions:

4.1 DEFINITIONS AND PROPERTY

Considering a linear ordinary differential equation of second order in the form

$$P(x)\frac{d^2 y}{dx^2} + Q(x)\frac{dy}{dx} + R(x)y = S(x), \tag{4.1}$$

we define

Definition 4.1.1

A singular point $x = x_0$, of equation (4.1), is said to be a 'regular singular point, (RSP)', if the following relations involving the coefficients P, Q and R hold good:

$$\left.\begin{aligned} &\text{(i)} \quad \lim_{x \to x_0} (x - x_0)\frac{Q(x)}{P(x)} = \text{Finite} \\ &\text{and} \\ &\text{(ii)} \quad \lim_{x \to x_0} (x - x_0)^2 \frac{R(x)}{P(x)} = \text{Finite} \end{aligned}\right\} \tag{4.2}$$

(Note that $P(x_0) = 0$, since $x = x_0$ is a singular point of equation (4.1)).

This definition can be easily generalized in cases of differential equations of higher order. But, as we are not going to be concerned in this book with higher order equations, with variable coefficients, than the second, we shall not consider such generalizations here.

If, therefore, $x = x_0$ is an RSP of equation (4.1), we can rewrite this equation, by the help of the relations (4.2), in the following general form:

$$(x-x_0)^2 \frac{d^2 y}{dx^2} + (x-x_0)\, q(x) \frac{dy}{dx} + r(x)\, y = s(x) \tag{4.3}$$

where $q(x)$ and $r(x)$ are analytic functions of x, in a certain neighbourhood of the singular point $x = x_0$, having the following Taylor expansions valid ther :

$$\left.\begin{aligned} q(x) &= \sum_{m=0}^{\infty} q_m (x-x_0)^m, \\ \text{and} \qquad r(x) &= \sum_{m=0}^{\infty} r_m (x-x_0)^m, \end{aligned}\right\} \tag{4.4}$$

with q_m and r_m as known constants.

The general solution of the inhomogeneous differential equation (4.3) can be determined completely, once the general solution of the corresponding homogeneous equation is determined, for, the method of variation of parameters described in §2.4 can then be employed in a straight-forward manner.

We are therefore motivated only to study the homogeneous differential equations of second order which is of the type

$$x^2 \frac{d^2 y}{dx^2} + xq(x) \frac{dy}{dx} + r(x)\, y = 0, \tag{4.5}$$

so far as solutions near regular singular points are concerned, where

$$\left.\begin{aligned} q(x) &= \sum_{m=0}^{\infty} q_m x^m, \\ \text{and} \qquad r(x) &= \sum_{m=0}^{\infty} r_m x^m. \end{aligned}\right\} \tag{4.6}$$

Note that equation (4.5) has an RSP at $x = 0$ and it takes care of regular singular points existing anywhere else, since a simple transformation of the variables reduces the left hand side of the general equation (4.3) into the left-hand side of the special equation (4.5).

We are now in a position to identify the following immediate property of a solution $y(x)$ of the differential equation (4.5) in a sufficiently small neighbourhood of the regular singular point $x = 0$:

Property

It is rather obvious that there exists a small neighbourhood of the point $x = 0$, in which, by using (4.6), the differential equation (4.5) can be approximated by the new equation as given by:

$$x^2 \frac{d^2 y}{dx^2} + q_0\, x \frac{dy}{dx} + r_0\, y = 0, \tag{4.7}$$

after assuming that $x^i y$ and $x^{i+1} y'$ are finite and small in the small neighbourhood under consideration, for all $i \geq 1$.

Now with q_0 and r_0 as known constants, equation (4.7) is an 'Euler-Cauchy' type of equation, considered earlier in section 2.2, and, therefore, the general solution of equation (4.7) has the form

$$y = c_1 x^{s_1} + c_2 x^{s_2}, \tag{4.8}$$

where s_1 and s_2 are the roots of the quadratic equation (see 2.2 for details):

$$s(s-1) + q_0 s + r_0 = 0, \tag{4.9}$$

in s, and c_1, c_2 are arbitrary constants.

This equation (4.8) has a special name in the theory of second-order ordinary differential equations as given below:

Definition 4.1.2

The quadratic equation (4.8) is called the 'indicial equation' corresponding to the second order ordinary differential equation (4.5), with the relations (4.6) holding good.

We are thus in a position to visualize that there exists a neighbourhood of the origin ($x = 0$) in which a solution $y(x)$ of the actual equation (4.5) behaves in such a manner that this solution $y(x)$ has the structure as given by

$$y(x) = x^s z(x, s) \tag{4.10}$$

where s is a root of the 'indicial equation' (4.8), and $z(x, s)$ represents an analytic function of x, for a given fixed value of s.

In fact, this important property, about the behaviour of one of the two solutions of equation (4.5) in the neighbourhood of the RSP $x = 0$, forms the basis of the famous method, known as Frobenius method, which will be described and utilized thoroughly in the remaining portion of this chapter.

4.2 THE FROBENIUS METHOD

The Frobenius method is the method of finding the general solution, of a second order linear homogeneous equation having a regular singular point at the origin. As observed in section 4.1, such an equation has the structure:

$$x^2 \frac{d^2 y}{dx^2} + xq(x) \frac{dy}{dx} + r(x) y = 0, \tag{4.11}$$

where

$$q(x) = \sum_{m=0}^{\infty} q_m x^m, \quad r(x) = \sum_{m=0}^{\infty} r_m x^m \tag{4.12}$$

and that one of the two solutions of this equation may be assumed in the form (4.10), i.e.,

$$y(x) = x^s z(x, s) \tag{4.13}$$

where s is a root of the 'indicial equation' (4.9), i.e.

$$s\,(s-1) + q_0\, s + r_0 = 0, \tag{4.14}$$

and

$$z\,(x, s) = \sum_{n=0}^{\infty} a_n\, x^n, \tag{4.15}$$

where the constants $a_n = a_n\,(s)$ will have to be determined.

The basis of Frobenius method is the assumption (4.13) on the form of a solution of equation (4.11), where s is a root of the indicial equation (4.14) and $z\,(x, s)$ is as given by equation (4.15), along with the further assumption that the first coefficient a_0 in the series for $z\,(x, s)$, *is not equal to zero* ($a_0 \neq 0$).

Under these basic assumptions, the next task in Frobenius method is to determine the remaining coefficients a_n ($n \geq 1$) of the series in equation (4.15), and this can be achieved by means of a procedure as described below.

We first rewrite equation (4.11) in the following alternative form, after using the relations (4.12):

$$x^2 \frac{d^2 y}{dx^2} + x \left(\sum_{m=0}^{\infty} q_m\, x^m \right) \frac{dy}{dx} + \left(\sum_{m=0}^{\infty} r_m\, x^m \right) y = 0. \tag{4.16}$$

Then, substituting (4.13) along with (4.15) in (4.16) [since (4.13) has been assumed to be a possible solution of equation (4.11)] we get

$$x^s \left[\sum_{n=0}^{\infty} a_n\,(s+n)\,(s+n-1)\, x^n + \sum_{n=0}^{\infty} (s+n)\, a_n \left(\sum_{m=0}^{\infty} q_m\, x^m \right) x^n \right.$$
$$\left. + \sum_{n=0}^{\infty} a_n \left(\sum_{m=0}^{\infty} r_m\, x^m \right) x^n \right] = 0. \tag{4.17}$$

Cancelling out the common factor x^s, this last equation can be cast into the form:

$$\sum_{n=0}^{\infty} a_n \left[(s+n)\,(s+n-1) + (s+n) \left(\sum_{m=0}^{\infty} q_m\, x^m \right) + \left(\sum_{m=0}^{\infty} r_m\, x^m \right) \right] x^n = 0 \tag{4.18}$$

and, this equation must be satisfied identically in some specified neighbourhood of the point $x = 0$. We are, therefore, led to the situation which demands that the coefficients of all powers of x on the left of equation (4.18) must vanish identically, and we obtain the following relations connecting the coefficients a_n:

"Coeff. of $x^0 = 0$" gives:

$$a_0\,[s\,(s-1) + sq_0 + r_0] = 0,$$

i.e. $$f\,(s) = s\,(s-1) + sq_0 + r_0 = 0 \tag{4.19}$$

(since $a_0 \neq 0$ (assumption)).

So, we get back the indicial equation (4.14) again.

"Coeff. of x^1 = 0" gives:

$$a_1 [(s+1)s + (s+1)q_0 + r_0] + a_0 [sq_1 + r_1] = 0,$$

i.e.,
$$a_1 = a_0 \frac{h_1(s)}{f(s+1)}, \tag{4.20}$$

where $h_1(s)$ is a polynomial in s $[h_1(s) = -(sq_1 + r_1)]$.

"Coeff. of x^2 = 0" gives:

$$a_2 [(s+2)(s+1) + (s+2)q_0 + r_0] + a_1 [(s+1)q_1 + r_1] + a_0 [(s+2)q_0 + r_0] = 0,$$

i.e.,
$$a_2 = \frac{-1}{f(s+2)} [a_1 \{(s+1)q_1 + r_1\} + a_0 \{(s+2)q_0 + r_0\}],$$

or

$$a_2 = a_0 \frac{h_2(s)}{f(s+1)f(s+2)}, \tag{4.21}$$

after utilizing the relation (4.20), where $h_2(s)$ is another polynomial in s, [We need not compute these polynomials $h_i(s)$ exactly, as will be clear later on].

The general recurrence relation for the coefficient a_n is obtained in a similar manner, by equating the coefficient of x^n to zero in the expression on the left of equation (4.18), and we find that such a recurrence relation has the structure

$$a_n = a_0 \frac{h_n(s)}{f(s+1)f(s+2)\ldots f(s+n)}, \quad (n \geq 1) \tag{4.22}$$

where $h_n(s)$ is a polynomial in s, the function $f(s)$ being the one as defined in equation (4.19).

We thus find that the unknown function $z(x, s)$, assumed in equation (4.15), is of the form

$$z(x, s) = a_0 x^s \left[1 + \sum_{n=1}^{\infty} \frac{h_n(s)\, x_n}{f(s+1)f(s+2)\ldots f(s+n)} \right], \tag{4.23}$$

if only the coefficients of all powers of x, except that of x^0 are equated to zero, in the expression on the left of equation (4.18), which suggest that $z(x, s)$ satisfies the equation

$$\left[x^2 \frac{d^2}{dx^2} + x\, q(x) \frac{d}{dx} + r(x) \right] z(x, s) = a_0 f(s) x^s, \tag{4.24}$$

If we also equate the coefficient of x^0 to zero in the expression in equation (4.18), we get the indicial equation (4.14), i.e. $f(s) = 0$, whose root (or roots) settles the value ((or values) of s to work with, and the relation (4.23) will then provide us with a solution (or both solutions) of the differential equation (4.11), if the coefficients

$$A_n(s) = \frac{h_n(s)}{f(s+1)\, f(s+2) \ldots f(s+n)}, \tag{4.25}$$

in the series on the right of equation (4.23) can be computed finitely for a root s (or for both the roots) of the indicial equation. We thus have the following theorem.

Theorem 4.1.1

In the case when the two roots s_1 and s_2 (or even one of the roots) of the indicial equation help the computation of the coefficients, $A_n(s_1)$ and $A_n(s_2)$ in finite forms, as given by equation (4.25), to be possible, we obtain the two possible solutions (or at least one solution) of the differential equation (4.11) in the forms [see equation (4.23)]:

$$y_1(x) = z(x, s_1) \quad \text{and} \quad y_2(x) = z(x, s_2),\} \tag{4.26}$$

But, as pointed out below, this refers to only very specific cases of equation (4.11) and we have to be extremely careful in cases of general interests.

From the form (4.25) of the coefficients A_n, it is clear that computational difficulties arise in the following two circumstances:

(i) When $f(s+i) = 0$, for a root s of the indicial equation $f(s) = 0$, and for some $i \geq 1$ and

(ii) When both $h_n(s)$ and $f(s+i)$ vanish for a root s of the indicial equation, and for some $i \geq 1$.

To be able to appreciate the above two difficulties which actually arise in practice, as will be seen later on through specific examples, we express the indicial equation.

$$f(s) \equiv s(s-1) + q_0 s + r_0 = 0$$

in the alternative form

$$f(s) \equiv (s - s_1)(s - s_2) = 0,$$

where s_1 and s_2 are the two roots of the indicial equation, assumed to be real for problems of interest to us.

Then we observe that

$$f(s+i) = (s + i - s_1)(s + i - s_2)$$

and that the difficulty $f(s+i) = 0$, for some i and for any one of the roots s_1 and s_2, never arises when $s_1 = s_2$ (equal roots).

Thus, the only way the equation $f(s+i) = 0$, for some i and for one of the roots s_1 and s_2 can be satisfied is that the roots s_1 and s_2 are real and distinct. We regard s_1 as the smaller root and s_2 the greater i.e., $s_2 > s_1$, without any loss of generality. Then we observe that $f(s_2+i)$ can never vanish, but $f(s+i) \equiv i\{i-(s_2-s_i)\}$ can vanish if $s_2 - s_1 = i$.

Hence, the factor that can be regarded to be common in the two difficulties (i) and (ii) as pointed out above, is that "the roots of the indicial equation are 'real' and 'distinct' and they differ by an integer". We also point out, at this stage, that Theorem 4.1.1, provides two independent solutions of equation (4.11), if s_1 and s_2 are real, and they do not differ by an integer.

The summary of the above observations is that there will exist computational difficulties for the coefficients A_n, as given by (4.25), when "The indicial equation $f(s) = 0$ has two real roots s_1 and s_2, $s_2 > s_1$ and $s_2 - s_1 = i$, for some i", and either, *Case A*; $h_i(s_1) \neq 0$, or *Case B*: $h_i(s_1) = 0$ holds.

In order to overcome the difficulty in *Case A*, we consider, the expression $(i = s_2 - s_1)$

$$f(s+i)\, z(x, s) = f(s+i)\,.\, a_0 x^s \left[1 + \sum_{n=1}^{\infty} A_n x^n\right]$$

$$= (s + 2s_2 - 2s_1)(s - s_1)\, a_0 x^s \left[1 + \sum_{n=1}^{\infty} A_n x^n\right].$$

We find that even though the coefficients $A_n(s_1)$ cannot be computed finitely for $n \geq i$, there exists no difficulty at all to compute the coefficients $(s - s_1) A_n$ for $s = s_1$ and thus the new function $f(s+i)\, z(x, s)$, or for that matter $(s - s_1)\, z(x, s)$ turns out to be a well-defined function of s for any fixed value of x, even for $s = s_1$.

There also exists no difficulty in seeing that this new function $(s - s_1) z(x, s)$ satisfies the equation [see (4.23) and (4.24)].

$$\left[x^2 \frac{d^2}{dx^2} + xq(x) \frac{d}{dx} + r(x)\right] \{(s - s_1)\, z(x, s)\} = a_0 f(s)(s - s_1)\, x^s. \quad (4.27)$$

If we now use the fact that $f(s) = (s - s_1)(s - s_2)$, we find, from equation (4.27), that the following theorem holds.

Theorem 4.1.2

The two functions

$$y_1(x) = \lim_{s \to s_1} [(s - s_1)\, z(x, s)] \quad (4.28)$$

and

$$y_2(x) = \lim_{s \to s_1} \frac{\partial}{\partial s} [(s - s_1)\, z(x, s)] \quad (4.29)$$

can be taken as two linearly independent solutions of the differential equation (4.11), in the case when the two roots s_1 and s_2 of the indicial equation are real and when they differ by an integer. The form of $z(x, s)$ in equations (4.28) and (4.29) is the one as given by equation (4.23).

While the first part of the proof of the above theorem follows from the arguments already explained, the part giving rise to 'linearly independent' nature of the two solutions becomes clear when one observes the following identity:

$$\frac{\partial}{\partial s}[(s-s_1)\,z\,(x,s)] = z\,(x,s) + (s-s_1)\frac{\partial z}{\partial s},$$

since $\frac{\partial z}{\partial s}$ will contain terms involving $\log x$, which is absent in $z\,(x, s)$, and hence follows the linear independence.

Note : As already pointed out before, the computation of the coefficients $A_n\,(s_2)$ for the larger root $s = s_2$ of the indicial equation causes no difficulty at all and, therefore, the function

$$y_3\,(x) = z\,(x, s_2) \tag{4.30}$$

can be taken to be a solution of the differential equation (4.11). But it so turns out (can be seen through examples at least !) that the function $y_3\,(x)$ and $y_1\,(x)$ differ from each other only by a multiplicative constant, and, therefore, y_3 is not linearly independent of y_1. Thus, the two independent solutions of equation (4.11), in Case *A* above, can always be taken to be $y_1\,(x)$ and $y_2\,(x)$ and, the general solution is given by

$$y\,(x) = c_1\,y_1\,(x) + c_2\,y_2\,(x),$$

where c_1, c_2 are two arbitrary constants, and y_1, y_2 are as given by equations (4.28) and (4.29).

Next, we shall consider the *Case B*, i.e., the case when $s_2 - s_1 = i$, as well as $h_i\,(s_1) = 0$.

In order to overcome the computational difficulty,as far as the computation of the coefficients $A_n\,(s_1)$ are concerned in this case, we can treat $A_i\,(s_1)$ to be a new arbitrary constant, since we have that

$$A_i\,(s) = \frac{h_i\,(s)}{f(s+1)\,f(s+2)\,\ldots f(s+i-1)\,(s+i-s_1)\,(s+i-s_2)}$$

$$= \frac{h_i\,(s)}{f(s+1)\,f(s+2)\,\ldots f(s+i-1)\,(s+s_2-2\,s_1)\,(s-s_1)}$$

and that $A_i\,(s_1)$ is of the form $\frac{0}{0}$ (indeterminate). With this artificial choice of the coefficient A_i, we notice from the recurrence relation (4.22), that all the coefficients a_n can be computed finitely for the root $s = s_1$ of the indicial equation, in terms of the two arbitrary constants a_0 and $a_0\,A_i\,(= a_i$, say), so that we obtain

$$z\,(x, s_1) = a_0\,\hat{y}_1\,(x) + a_i\,\hat{y}_2\,(x), \tag{4.31}$$

where $\hat{y}_1(x)$ and $\hat{y}_2(x)$ are the two series that come out from the assumption (4.15) of the solution of the differential equation (4.11), and these two functions $\hat{y}_1$ and $\hat{y}_2$ provide us with the two linearly independent solutions of the given differential equation (4.11) [linearly independent because, the structures of $\hat{y}_1$ and $\hat{y}_2$ are:

$$\hat{y}_1(x) = x^{s_1}[1 + (\ldots)x + (\ldots)x^2 + \ldots]$$

and

$$\hat{y}_2(x) = x^{s_1 + i(\equiv s_2)}[1 + (\ldots)x + (\ldots)x^2 + \ldots]$$

and

$$s_2 > s_1].$$

Note : As in case *A*, the function $z(x, s_2)$ in this case *B* also provides us with a solution of the differential equation (4.11), but this new solution can be shown to be a constant multiple of the solution $\hat{y}_2(x)$.

We have, thus, been able to sort out the main difficulties existing in the computation of the coefficients of the series $z(x, s)$ in the above two cases *A* and *B*, when the roots of the indicial equation $f(s) = 0$ are real and when they differ by an integer, and have devised a method of obtaining the two linearly independent solutions of the given differential equation (4.11) in such circumstances.

What remains next, in the Frobenius method, to be done is that we have to have a device by the help of which we can obtain the two linealy independent solutions of equation (4.11), in the case when the two roots of the indicial equation are equal, i.e., $s_1 = s_2$.

As pointed out before, the expression $z(x, s_1)$ (cf. equation (4.23)) provides us with only one solution of the differential equation (4.11), and we write this solution as

$$y_1(x) = z(x, s_1) \tag{4.32}$$

In order to obtain the second solution, in this case of equal roots of the indicial equation, we go back to equation (4.24) and obtain

$$\left[x^2 \frac{d^2}{dx^2} + xq(x)\frac{d}{dx} + r(x)\right] z(x, s) = a_0 (s - s_1)^2 x^s. \tag{4.33}$$

We thus find that the function

$$y_2(x) = \lim_{s \to s_1} \left\{\frac{\partial}{\partial s} z(x, s)\right\} \tag{4.34}$$

provides us with the second linearly independent solution of the differential equation (4.11).

We summarise the above results in the form of the following theorem:

Theorem 4.1.3

The two functions

$$y_1(x) = \lim_{s \to s_1} z(x, s) = z(x, s_1)$$

and

$$y_2(x) = \lim_{s \to s_1} \frac{\partial z(x, s)}{\partial s}$$

can be taken to be two linearly independent solutions of the differential equation (4.11), in the case when the indicial equation possesses two equal roots.

The theory and methods described above are collectively known as the 'Frobenius' method and the real appreciation of this method only comes when concrete examples are taken up and differential equations are solved by this method successfully. The remaining part of this chapter will, therefore, be devoted to solving specific examples by the method of Frobenius.

4.2.1 Examples

In this section we shall consider four typical examples of second order linear ordinary differential equations having a regular singular point at the origin (i.e., $x = 0$) and show how to obtain both the solutions of each of these differential equations by 'Frobenius' method. Some more examples are provided in the later chapters (Ch. 6, 7, 8) by way of treating some of the very important differential equations giving rise to special functions.

Example (i)

$$4x \frac{d^2 y}{dx^2} + \frac{dy}{dx} - y = 0. \tag{i.1}$$

The 'indicial equation' for this differential equation can be written down immediately after rewriting it as

$$x^2 \frac{d^2 y}{dx^2} + x \cdot \frac{1}{4} \frac{dy}{dx} - \frac{x}{4} y = 0 \tag{i.2}$$

and observing that, in this case, $q_0 = 1/4$ and $r_0 = 0$.

We thus find that the 'indicial equation' is given by

$$f(s) \equiv s(s-1) + \frac{1}{4} \cdot s \equiv s\left(s - \frac{3}{4}\right) = 0, \tag{i.3}$$

giving the two roots

$$s_1 = 0 \text{ and } s_2 = \frac{3}{4}. \tag{i.4}$$

The case of this particular example, is, therefore, a straightforward one, and we find that the two solutions can be written down as

$$y_1(x) = z(x, s_1) \text{ and } y_2(x) = z(x, s_2) \tag{i.5}$$

where

$$z(x, s) = \sum_{n=0}^{\infty} a_n x^{n+s} \; (a_0 \neq 0), \tag{i.6}$$

and the coefficients a_n $(n \geq 1)$ are to be determined as explained in the general theory described in the previous section.

Substituting $y = z(x, s)$, as given by (i.6), into (i.2) we get (after cancelling out the factors x^s):

$$\sum_{n=0}^{\infty} a_n x^n \left[(n+s)(n+s-1) + \frac{1}{4}(n+s) - \frac{x}{4} \right] = 0$$

or

$$\sum_{n=0}^{\infty} a_n x^n \left[(n+s)\left(n+s-\frac{3}{4}\right) - \frac{x}{4} \right] = 0. \tag{i.7}$$

Equating coefficients of x^n to zero we get the recurrence relation

$$a_n = + \frac{a_{n-1}}{4(n+s)\left(n+s-\frac{3}{4}\right)}, \quad (n \geq 1) \tag{i.8}$$

and repeated use of this relation gives rise to the following results:

$$\left.\begin{aligned} a_1 &= + \frac{a_0}{4(s+1)\left(s+\frac{1}{4}\right)}, \\ a_2 &= + \frac{a_0}{4^2 (s+1)(s+2)\left(s+\frac{1}{4}\right)\left(s+\frac{5}{4}\right)}, \\ a_3 &= + \frac{a_0}{4^3 (s+1)(s+2)(s+3)\left(s+\frac{1}{4}\right)\left(s+\frac{5}{4}\right)\left(s+\frac{9}{4}\right)}, \\ a_n &= \frac{a_0}{4^n (s+1)(s+2)\ldots(s+n)\left(s+\frac{1}{4}\right)\left(s+\frac{5}{4}\right)\left(s+\frac{9}{4}\right)\ldots\left(s+n-\frac{3}{4}\right)}. \end{aligned}\right\} \tag{i.9}$$

Putting $s = s_1 = 0$ and $s = s_2 = \frac{3}{4}$, in succession, we get,

$$\left.\begin{aligned} a_1(s_1) &= \frac{a_0}{(1\,!)}, \quad a_2(s_1) = \frac{a_0}{(2\,!)(1.5)} \\ a_3(s_1) &= \frac{a_0}{(3\,!)(1.5.9)}, \ldots, \quad a_n(s_1) = \frac{a_0}{(n\,!)\,[1.5.9\ldots(4n-3)]} \end{aligned}\right\} \tag{i.10}$$

and

$$\left.\begin{aligned} a_1(s_2) &= \frac{a_0}{1\,.\,1\,!\,7}, \quad a_2(s_2) = \frac{a_0}{1\,.\,(2\,!)\,(7.11)}, \\ a_3(s_2) &= \frac{a_0}{1\,.\,(3\,!)\,(7.11.15)}, \\ &\ldots\ldots, \\ a_n(s_2) &= \frac{a_0}{1\,.\,(n\,!)\,[7.11.15\ldots(4n+3)]}. \end{aligned}\right\} \tag{i.11}$$

The two independent solutions of the differential equation (i.1) are then obtained in the forms:

$$y_1(x) = z(x, s_1) = x^{s_1} \sum_{n=0}^{\infty} a_n(s_1)\, x^n \equiv \sum_{n=0}^{\infty} a_n(s_1)\, x^n,$$

$$= a_0 \left[1 + \sum_{n=0}^{\infty} \frac{x^n}{(n\,!)\,\{1.5.9\ldots(4n-3)\}} \right], \tag{i.12}$$

and

$$y_2(x) = z(x, s_2) = x^{s_2} \sum_{n=0}^{\infty} a_n(s_2)\, x^n \equiv x^{3/4} \sum_{n=0}^{\infty} a_n(s_2)\, x^n$$

$$= a_0\, x^{3/4} \left[1 + \sum_{n=1}^{\infty} \frac{x^n}{(n\,!)\,\{7.11.5\ldots(4n+3)\}} \right], \tag{i.13}$$

where a_0 is an arbitrary constant.

Example (ii)

$$x \frac{d^2 y}{dx^2} - 2 \frac{dy}{dx} + y = 0. \tag{ii.1}$$

Rewriting the given equation in the form

$$x^2 \frac{d^2 y}{dx^2} - 2x \frac{dy}{dx} + xy = 0, \tag{ii.2}$$

we find that $q_0 = -2$ and $r_0 = 0$, so that the 'indicial equation' is given by

$$f(s) \equiv s(s-3) = 0, \tag{ii.3}$$

having the two real roots $s_1 = 0$ and $s_2 = 3$ and, therefore $s_2 - s_1 = 3 =$ a positive integer.

Thus, in this case, we are going to face computational problems and the situation can be either of the two cases *A* or *B* described earlier. In order to identify the actual difficulty we substitute

$$y = z(x, s) = \sum_{n=0}^{\infty} a_n\, x^{n+s},$$

in the given equation (ii.1) and find that we must have the identity

$$\sum_{n=0}^{\infty} a_n x^n [(n+s)(n+s-3)+x] = 0, \tag{ii.4}$$

from which results the recurrence relation

$$a_n = -\frac{a_{n-1}}{(n+s)(n+s-3)}, \quad (n \geq 1) \tag{ii.5}$$

or

$$a_n = \frac{(-1)^n a_0}{\{(s+1)(s+2)\dots(s+n)\}\,.\,\{(s-2)(s-1)s(s+1)\dots(s+n-3)\}} \quad (n \geq 1) \tag{ii.6}$$

and therefore, in this example, $h_n(s) = (-1)^n$, in the notations of equation (4.22) and (4.25).

Having thus identified that the present problem falls under *Case A*, *Theorem* 4.1.2, along with equation (4.23) decides that the two solutions are given by

$$y_1(x) = \lim_{s \to s_1 (=0)} [(s-s_1)\, z(x,s)], \tag{ii.7}$$

and

$$y_2(x) = \lim_{s \to s_1 (=0)} \frac{\partial}{\partial s} [(s-s_1)\, z(x,s)], \tag{ii.8}$$

where

$$z(x,s) = a_0 x^s \left[1 + \sum_{n=1}^{\infty} \frac{(-1)^n x^n}{\{(s+1)(s+2)\dots(s+n)\}\{(s-2)(s-1)s(s+1)\dots(s+n-3)\}}\right] \tag{ii.9}$$

We find that we easily obtain

$$y_1(x) = a_0 \left[\frac{x}{2} + \frac{x^2}{4} + \sum_{n=3}^{\infty} \frac{(-1)^n x^n}{2\,(n!)\,(n-3)!}\right] \quad (0! = 1) \tag{ii.10}$$

and

$$y_2(x) = a_0 \lim_{s \to 0} \frac{\partial}{\partial s}\left[s x^s + x^s \sum_{n=1}^{\infty} \frac{(-1)^n x^n}{\{(s+1)(s+2)\dots(s+n)\},\{(s-2),(s-1)(s+1)\dots(s+n-3)\}}\right]$$

$$= a_0 + y_1(x)\log x + a_0 \sum_{n=1}^{\infty} (-1)^n x^n \left(\frac{d\lambda_n(s)}{ds}\right)_{s=0} \tag{ii.11}$$

where

$$\lambda_n(s) = \frac{1}{\{(s+1)(s+2)\ldots(s+n)\}.\{(s-2).(s-1)(s+1)\ldots(s+n-3)\}}. \tag{ii.12}$$

For the computation of the quantities $(d\lambda_n/ds)_{s=0}$, we proceed as follows: We have the result that

$$\frac{d}{ds}(\log \lambda_n) = \frac{1}{\lambda_n}\frac{d\lambda_n}{ds},$$

$$\therefore \quad \frac{d\lambda_n}{ds} = \lambda_n \frac{d}{ds}(\log \lambda_n).$$

Then (ii.12) gives that

$$\left(\frac{d\lambda_n}{ds}\right)_{s=0} = \frac{-1}{n!\{(-2).(-1).1.2\ldots(n-3)\}}\left[\frac{1}{1}+\frac{1}{2}+\frac{1}{3}+\ldots\frac{1}{n}+\frac{1}{-2}\right.$$
$$\left.+\frac{1}{-1}+\frac{1}{1}+\frac{1}{2}+\frac{1}{3}+\ldots+\frac{1}{n-3}\right],$$

giving

$$\left(\frac{d\lambda_1}{ds}\right)_{s=0} = +\frac{1}{2}.0 = 0$$

$$\left(\frac{d\lambda_2}{ds}\right)_{s=0} = +\frac{1}{2.2!}\left[1+\frac{1}{2}-1+1\right] = \frac{3}{8},$$

and

$$\left(\frac{d\lambda_n}{ds}\right)_{s=0} = \frac{1}{2n!(n-3)}\left[1+2\left(\frac{1}{2}+\frac{1}{3}+\ldots+\frac{1}{n-3}\right)\right.$$
$$\left.+\frac{1}{n-2}+\frac{1}{n-1}+\frac{1}{n}\right], \quad (n \geq 3). \tag{ii.13}$$

Substituting from (ii.13) into (ii.11) we finally obtain the appropriate form of the second solution $y_2(x)$ of the given differential equation and its linear independence from the first solution $y_1(x)$ is also seen very easily.

Example (iii)

$$x^2\frac{d^2y}{dx^2} + x\frac{dy}{dx} + \left(x^2 - \frac{1}{4}\right)y = 0. \tag{iii.1}$$

The 'indicial equation' for this example is given by

$$f(s) \equiv s^2 - \frac{1}{4} = 0, \tag{iii.2}$$

having the two roots $s_1 = -\frac{1}{2}$, $s_2 = \frac{1}{2}$, so that $s_2 - s_1 = 1 = a$ positive integer.

The actual substitution of

$$y(x) = z(x, s) = \sum_{n=0}^{\infty} a_n x^{n+s} \qquad \text{(iii.3)}$$

into the given equation (iii.1) produces the identity

$$\sum_{n=0}^{\infty} a_n x^n \left[\left\{(n+s)^2 - \frac{1}{4}\right\} + x^2\right] = 0, \qquad \text{(iii.4)}$$

giving the recurrence relations:

$$a_1 \left[(s+1)^2 - \frac{1}{4}\right] = 0, \qquad \text{(iii.5)}$$

and

$$a_n \left[(s+n)^2 - \frac{1}{4}\right] = -\, a_{n-2}, \; (n \geq 2), \qquad \text{(iii.6)}$$

The fact that $s_2 - s_1 = 1$ and that the relation (iii.4) holds good for the coefficient a_1 (cf. eqn. (4.22)), in this particular example, suggest that this example falls under *Case B* of the general theory and hence the two solutions of the differential equation can be obtained from the single representation (iii.3), if we treat a_1 as another arbitrary constant, together with the root $s_1 = -\dfrac{1}{2}$ of the indicial equation.

We then find, from the recurrence relation (iii.6), that

$$a_n = -\frac{a_{n-2}}{\left(s + n - \dfrac{1}{2}\right)\left(s + n + \dfrac{1}{2}\right)}, \; n \geq 2$$

so that we have

$$\left.\begin{aligned} a_2(s_1) &= -\frac{a_0}{1.2} = -\frac{a_0}{2!}, \\ a_3(s_1) &= -\frac{a_1}{2.3} = -\frac{a_1}{3!}, \\ a_4(s_1) &= -\frac{a_2(s_1)}{3.4} = +\frac{a_0}{4\,!}, \\ a_5(s_1) &= -\frac{a_3(s_1)}{4.5} = +\frac{a_1}{5\,!}, \end{aligned}\right\} \qquad \text{(iii.7)}$$

and so on.

Substituting from (iii.7) into (iii.3) and using the result (4.31), we find that the two solutions of the given differential function equation (iii.1) are given by:

$$\left.\begin{aligned} \hat{y}_1(x) &= x^{-1/2}\left[1 - \frac{x^2}{2\,!} + \frac{x^4}{4\,!} - \dots\right] \\ &\equiv x^{-1/2}\cos x, \\ \text{and} \\ \hat{y}_2(x) &= x^{-1/2+1}\left[1 - \frac{x^2}{3\,!} + \frac{x^4}{5\,!} - \dots\right] \\ &= x^{-1/2}\sin x. \end{aligned}\right\} \quad \text{(iii.8)}$$

It can be easily checked, by direct substitutions, that the above two solutions $\hat{y}_1$ and $\hat{y}_2$ do really satisfy the given differential equation (iii.1) and that they also represent two linearly independent solutions can be verified by calculating the 'Wronskian' which is $1/x$.

Example (iv)

$$x\frac{d^2y}{dx^2} + (x+1)\frac{dy}{dx} + y = 0. \quad \text{(iv.1)}$$

The 'indicial equation' here is given by

$$f(s) \equiv s^2 = 0, \quad \text{(iv.2)}$$

having two 'equal' roots $s = 0$.

The two solutions for this differential equaiton will be obtained by using *Theorem* 4.1.3, with the help of the substitution

$$y(x) = z(x, s) = \sum_{n=0}^{\infty} a_n x^{n+s} \quad \text{(iv.3)}$$

into the given equation (iv.1), which gives rise to the identity

$$\sum_{n=0}^{\infty} a_n x^n \left[(n+s)^2 + (n+s+1)\,x\right] = 0, \quad \text{(iv.4)}$$

and the recurrence relation

$$a_n = -\frac{(n+s)}{(n+s)^2}\,a_{n-1},\ (n \geq 1) \quad \text{(iv.5)}$$

i.e.,

$$a_n = (-1)^n \frac{1}{(s+1)(s+2)\dots(s+n)}\,a_0,\ \ (n \geq 1). \quad \text{(iv.6)}$$

We thus obtain, by using (iv.6) in (iv.3), that

$$z(x, s) = a_0 \sum_{n=0}^{\infty} (-1)^n \frac{x^{n+s}}{(s+1)(s+2)(s+3)\dots(s+n)} \quad \text{(iv.7)}$$

and *Theorem* 4.1.3 gives the two solutions of equation (iv.1) in the forms:

$$y_1(x) = z(x, s_1) \equiv z(x, 0)$$

$$= a_0 \sum_{n=0}^{\infty} \frac{(-1)^n}{n!} x^n, \qquad \text{(iv.8)}$$

and

$$y_2(x) = \lim_{\varepsilon \to s_1 (=0)} \frac{\partial z(x, s)}{\partial s}$$

$$= y_1(x) \log x + a_0 \sum_{n=1}^{\infty} (-1)^n \left(\frac{d\lambda_n}{ds}\right)_{s=0} x^n, \qquad \text{(iv.9)}$$

where

$$\lambda_n(s) = \frac{1}{(s+1)(s+2)(s+3)\ldots(s+n)},$$

giving, as described in example (ii).

$$\left(\frac{d\lambda_n}{ds}\right)_{s=0} = \left(-\frac{2}{1} - \frac{1}{2} - \frac{1}{3} \cdots - \frac{1}{n}\right)\frac{1}{(n!)} \qquad \text{(iv.10)}$$

The linear independence of the two solutions y_1 and y_2, in this example is also obvious.

PROBLEMS

1. Find the general solutions of the following equations in the neighbourhood of the RSP mentioned within the brackets: ($y' = dy/dx$, etc).
   (i) $2xy'' + y' + y = 0$, $(x = 0)$
   (ii) $x^2 y'' - 3xy' + (3 - x)y = 0$, $(x = 0)$
   (iii) $x^2 y'' + (x^2 - 3x)y' + (4 - 2x)y = 0$, $(x = 0)$
   (iv) $x^2 y'' + xy' + (x^2 - 16)y = 0$, $(x = 0)$
2. (a) Find a solution $\phi(x)$ of the form

$$\phi(x) = (x-1)^r \sum_{k=0}^{\infty} c_k (x-1)^k$$

   for the 'Legendre' equation (see Chapter 5).

$$(1 - x^2)y'' - 2xy' + n(n+1)y = 0.$$

   For what values of x does the series converge ?

   (b) Show that there is a polynomial solution if n is a non-negative integer.
3. The equation

$$xy'' + (1 - x)y' + \nu y = 0$$

where ν is a constant, is called 'Laguerre' equation (see Chapter 6).

(a) Show that this equation has a regular singular point at $x = 0$, and find the general solution of this equation by using Frobenius' method.

(b) Show that if $\nu = n$, a non-negative integer, there is a polynomial solution of degree n.

4. The equation

$$x^2 y'' + xy' + (x^2 - n^2) y = 0,$$

where n is a constant, is known as 'Bessel' equation (see Chapter 7). Using Frobenius' method, obtain the general solution of Bessel equation for all possible values of n.

5. The equation

$$x(1-x)\frac{d^2 y}{dx^2} + \{\gamma - (\alpha + \beta + 1)x\}\frac{dy}{dx} - \alpha\beta y = 0$$

(α, β, γ are constants)

is called the 'Hypergeometric Equation'. Show that $x = 0$ and $x = 1$ are 'regular singular points' of the above equation, with the two roots of the 'indicial equation' corresponding to the points $x = 0$ and $x = 1$ as given by $0, 1 - \gamma$ and $0, \gamma - \alpha - \beta$ respectively.

By using the transformation $x = 1/t$ and letting $t \to 0$ afterwards, show that the point at 'infinity', *i.e.*, $x = \infty$ is also a regular singular point of the Hypergeometric Equation, with the two roots of the 'indicial equation' as α and β.

Hence obtain the series solutions valid in a neighbourhood of each of the regular singular points, by using Frobenius method.

6. Solve the equation

$$(1 - x^2)\frac{d^2 y}{dx^2} - (\alpha + \beta + 1)\frac{dy}{dx} - \alpha\beta y = 0$$

in a neighbourhood of $x = 1$, if α and β are real constants. What happens to the solution if $\alpha + \beta$ is not an odd integer ?

Deduce the general solution of the equation

$$\frac{d^2 y}{d\theta^2} + (\alpha + \beta)\cot\theta\frac{dy}{d\theta} - \alpha\beta y = 0$$

where $0 < \theta < \pi$ and α, β are real constants, such that $\alpha + \beta$ is not an odd integer.

(Hint: set $x = \cos\theta$).

7. The equation

$$\frac{d^2 y}{dx^2} + \frac{2}{x}\frac{dy}{dx} + \left(-\frac{1}{4} + \frac{k}{x} - \frac{l(l+1)}{x^2}\right) y = 0,$$

where k is a constant and l is zero or a positive integer, is known as Schrodinger's equation for the hydrogen atom (see Jaeger [7]); show that the above equation can be transformed to the equation

$$\frac{d^2 v}{dx^2} + \left(\frac{2}{x} - 1\right)\frac{dv}{dx} + \left\{\frac{k-1}{x} - \frac{l(l+1)}{x^2}\right\} v = 0,$$

by the transformation $y = v e^{-x/2}$.
Hence, show that a solution of the second equation can be determined in the form

$$v = x^s \sum_{n=0}^{\infty} a_n x^n,$$

where the 'indicial equation' satisfied by s has the two roots $s = l$ and $s = -l - 1$ and the coefficients a_n, for the root $s = l$ of the indicial equation satisfy the recurrence relation.

$$a_n = \frac{l+n-k}{n(n+2l+1)} a_{n-1}, \quad (n = 1, 2, 3, \ldots)$$

[*Note* : If l is zero or a positive integer, the recurrence relation above shows that all the coefficients a_n onwards vanish for $k = l + n$, $(n = 1, 2, 3, \ldots)$, so that v becomes a polynomial in x, of degree $l + n - 1$].

CHAPTER - FIVE

The Legendre Functions

In this chapter we shall be concerned with two important special functions known as the Legendre functions, which represent the two linearly independent solutions of Legendre's differential equation, under certain special circumstances.

We start our discussion on these two special functions, in a manner that is slightly different from that employed in the study of Hermite's functions taken up in Chapter 3.

5.1 THE LEGENDRE POLYNOMIAL

5.1.1 Basic Definition and Important Relations

Definition 5.1.1

The polynomial of degree n, defined by the relation

$$P_n(x) = \frac{1}{2^n} \sum_{i=0}^{m} \frac{(-1)^r (2n-2r)!}{r!\,(n-r)!\,(n-2r)!} x^{n-2r}, \tag{5.1}$$

where $m = n/2$ or $(n-1)/2$, according as n is even or odd, is known as Legendre's polynomial of degree n.

We immediately prove the following important results on Legendre polynomials.

Theorem 5.1.1

The following 'Rodrigue's' formula for Legendre Polynomial holds good:

$$P_n(x) = \frac{1}{2^n\,n!} \frac{d^n}{dx^n} \{(x^2-1)^n\} \tag{5.2}$$

Proof

Use the familiar binomial expansion

$$(x^2-1)^n = \sum_{k=0}^{n} \frac{(-1)^k\,n!}{k!\,(n-k)!} x^{2n-2k} \tag{5.3}$$

and the rest of the proof is straightforward.

Application.

Using Leibnitz's rule of differentiation of a product of two functions, we easily deduce that $P_n(1) = 1$ and $P_n(-1) = (-1)^n$.

Theorem 5.1.2

The function $w(x, t) = (1 - 2xt + t^2)^{-1/2}$ is the 'generating function' of the Legendre polynomials, in the sense that the expansion

$$w(x, t) = \sum_{n=0}^{\infty} P_n(x)\, t^n \tag{5.4}$$

holds for sufficiently small values of $|t|$.

Proof

Let r_1 and r_2 denote the two roots of the quadratic equation

$$1 - 2xt + t^2 = 0. \tag{5.5}$$

Then using a familiar theorem on binomial expansion, we can expand the function $w(x, t)$ as follows:

$$w(x, t) = [1 - (2tx - t^2)]^{-1/2} = \sum_{k=0}^{\infty} \frac{(1/2)_k}{k!} (2tx - t^2)^k, \tag{5.6}$$

the expansion being valid for $|t| < r = \min(|r_1|, |r_2|)$, where the symbol $(p)_r$ represents the product

$$(p)_r = p(p+1)(p+2)\ldots(p+r-1) = \frac{\Gamma(p+r)}{\Gamma(p)}. \tag{5.7}$$

Expanding binomially the expression $(2tx - t^2)^k$, appearing in equation (5.6), we obtain

$$w(x, t) = \sum_{k=0}^{\infty} \frac{(1/2)_k}{k!} t^k \sum_{s=0}^{k} (-1)^{k-s} \frac{k!}{s!\,(k-s)!} (2x)^s\, t^{k-s}. \tag{5.8}$$

By employing the convention that $(k-s)! \to \infty$ for $s > k$, we can express the result (5.8) in the form:

$$w(x, t) = \sum_{k=0}^{\infty} \sum_{s=0}^{\infty} \frac{(1/2)_k (-1)^{k-s}}{s!\,(k-s)!} (2x)^s\, t^{2k-s}. \tag{5.9}$$

The double series on the right hand side of equation (5.9) can be expressed in the form

$$\sum_{n=0}^{\infty} F_n(x)\, t^n, \tag{5.10}$$

where, it can be easily found after rearranging (*) the double series properly, that

(*) [we rewrite (5.9) as:

$$w(x, t) = \sum_{s=0}^{\infty} \sum_{k=0}^{\infty} (\ldots) = \sum_{s=0}^{\infty} \sum_{k=s}^{\infty} (\ldots) \quad (\because (k-s)! \to \infty \text{ for } k < s)$$

$$= \sum_{s=0}^{\infty} \sum_{k'=0}^{\infty} \frac{(1/2)_{k'+s} (-1)^{k'}}{s!\,k'!} (2x)^s \cdot t^{2k'+s} \quad (k - s = k').$$

Now, $2k' + s = n \Rightarrow s = n - 2k'$; also $q! \to \infty$ if q is a negative integer. Then, changing k' to p, the result (5.11) follows.]

$$F_n(x) = \sum_{p=0}^{m} \frac{(-1)^p (1/2)_{n-p}}{P!\,(n-2p)!} (2x)^{n-2p}, \tag{5.11}$$

with $m = n/2$ or $\frac{n-1}{2}$, according as n is even or odd. If we next use the result

$$\left(\frac{1}{2}\right)_n = \frac{\Gamma(1/2+n)}{\Gamma(1/2)} = \frac{1}{2^{2n}} \cdot \frac{(2n)!}{n!}, \tag{5.12}$$

we find that

$$F_n(x) = P_n(x) \tag{5.13}$$

and our theorem is proved for those values of t for which $|t| < r$, as mentioned earlier.

Definition 5.1.2

The Legendre Polynomial $P_n(x)$ is also called the Legendre function of the first kind and of positive integral order 'n'. For the general definition of Legendre functions of first kind for any constant value of n, real or complex, the reader is referred to McRobert's book [10].

We now prove the following 'Recurrence relations'.

Theorem 5.1.3

The following two recurrence relations hold for Legendre polynomials $P_n(x)$.

(i) $P'_{n+1}(x) - 2xP'_n(x) + P'_{n-1}(x) - P_n(x) = 0$

and

(ii) $(n+1)P_{n+1}(x) - (2n+1)xP_n(x) + nP_{n-1}(x) = 0$, for $n = 1, 2, 3, \ldots$

Proof

Differentiating the function $w(x, t)$ of *Theorem* 5.1.3 with respect to x and t respectively we obtain the following two identities:

$$\left.\begin{aligned} &\text{(i)}\quad (1-2xt+t^2)\frac{\partial w}{\partial x} - tw = 0 \\ &\text{and} \\ &\text{(ii)}\quad (1-2xt+t^2)\frac{\partial w}{\partial t} + (t-x)\,w = 0. \end{aligned}\right\} \tag{5.14}$$

Using the result of *Theorem* 5.1.3, i.e.

$$w(x, t) = \sum_{n=0}^{\infty} P_n(x)\, t^n,$$

in the above two equations (5.14) and equating the coefficients of t^n from both sides of the resulting identities, the recurrence relations (i) and (ii) follow immediately.

Deductions

1. Differentiating the relation (ii) with respect to x and eliminating $P'_{n+1}(x)$ between the resulting relation and relation (i), we get

(iii) $xP'_n(x) - P'_{n-1}(x) = nP_n(x).$

2. Eliminating $xP'_n(x)$ between the relation (iii) and the differentiated relation (ii), we obtain

(iv) $P'_{n+1}(x) - P'_{n-1}(x) = (2n+1)P_n(x).$

3. Eliminating $P'_{n-1}(x)$ between the two relations (iii) and (iv), we obtain

(v) $P'_{n+1}(x) - xP'_n(x) = (n+1)P_n(x).$

4. We also find that

$$\begin{aligned}(1-x^2)P'_n(x) &= P'_n(x) - x \cdot xP'_n(x)\\ &= P'_n(x) - x(nP_n(x) + P'_{n-1}(x)),\end{aligned}$$

by using (iii).

or,

$$(1-x^2)P'_n(x) = P'_n(x) - nxP_n(x) - (P'_n(x) - nP_{n-1}(x))$$

(by using (v)).

We thus obtain the result that

(vi) $(1-x^2)P'_n(x) = nP_{n-1}(x) - nxP_n(x).$

5. Differentiating the relation (vi) *w.r.t. x*, we get,

$$\begin{aligned}\frac{d}{dx}[(1-x^2)P'_n(x)] &= nP'_{n-1}(x) - nP_n(x) - nxP'_n(x)\\ &= -nP_n(x) + n(-nP_n(x)),\end{aligned}$$

by using (iii).

We thus find that $P_n(x)$ satisfies the O.D.E.

(vii) $$\frac{d}{dx}\left[(1-x^2)\frac{dP_n}{dx}\right] + n(n+1)P_n = 0, \qquad (5.15)$$

and this is the famous Legendre equation (see Problem 2, Chapter 4), in which n is a non-negative integer.

5.1.2 Some Important Properties

(A) The Orthogonality Property

Theorem 5.1.4

The set of Legendre Polynomials $\{P_n(x)\}$, $n = 0, 1, 2, \ldots$ satisfy the Orthogonality relations as given by

$$\int_{-1}^{1} P_m(x)P_n(x)\,dx = \frac{2}{2n+1}\delta mn,$$

in the interval $(-1, 1)$.

Proof

We shall first evaluate the integral

$$I_{m,n} = \int_{-1}^{1} x^m P_n(x)\,dx, \tag{5.16}$$

for all non-negative integers m and n.

Using Rodrigue's formula (*Theorem* 5.1.1) we find that

$$\int_{-1}^{1} f(x)\,P_n(x)\,dx = \frac{1}{2^n\,n\,!}\int_{-1}^{1} f(x)\frac{d^n}{dx^n}\{(x^2-1)^n\}\,dx$$

$$= \frac{(-1)^n}{2^n\,n\,!}\int_{-1}^{1} f^{(n)}(x)\cdot(x^2-1)^n\,dx; \tag{5.17}$$

(integrating by parts)

where $f(x)$ is an n-times differential function of x, in the interval $(-1, 1)$. Using the result (5.17) in (5.16) we immediately obtain that

$$I_{m,n} = \begin{cases} = 0, \text{ if } m < n \\ = \dfrac{(-1)^n}{2^n\,n\,!}\,\dfrac{m\,!}{(m-n)\,!}\displaystyle\int_{-1}^{1} x^{m-n}(x^2-1)^n\,dx, \text{ if } m > n. \end{cases} \tag{5.18}$$

We further find that, in the case when $m > n$, we have

$$I_{m,n} = \begin{cases} 0, \text{ if } m-n \text{ is odd} \\ \dfrac{m\,!}{2^{n-1}\,n\,!\,(m-n)\,!}\displaystyle\int_{-1}^{1} x^{m-n}(1-x^2)^n\,dx, \text{ if } m-n \text{ is even.} \end{cases} \tag{5.19}$$

Next, using the result

$$\int_0^1 y^{\frac{m-n-1}{2}}(1-y)^n\,dy = B\left(\frac{m-n+1}{2},\, n+1\right)$$

$$= \frac{\Gamma\left(\dfrac{m-n+1}{2}\right)\Gamma(n+1)}{\Gamma\left(\dfrac{m+n+3}{2}\right)}, \tag{5.20}$$

where $B(m, n)$ is the familiar Beta-function, we obtain, from (5.18) and (5.19), the following final result:

$$I_{m,n} = \begin{cases} 0, \text{ if } m < n \\ 0, \text{ if } m > n, \text{ with } m-n \text{ odd} \\ \dfrac{m\,!}{2^n\,(m-n)\,!} \dfrac{\Gamma\left(\dfrac{m-n+1}{2}\right)}{\Gamma\left(\dfrac{m+n+3}{2}\right)}, \text{ if } m > n, \text{ with } m-n \text{ even.} \end{cases} \tag{5.21}$$

We also obtain, by using the result (5.17), that

$$I_n \equiv I_{n,n} = \frac{(-1)^n}{2^n} \int_{-1}^{1} (x^2-1)^n\,dx = \frac{1}{2^{n-1}} \int_0^1 (1-x^2)^n\,dx$$

$$= \frac{1}{2^n} \int_0^1 (1-y)^n\,y^{-1/2}\,dy \quad (y = x^2)$$

$$= \frac{1}{2^n} B\left(n+1, \frac{1}{2}\right) = \frac{1}{2^n} \frac{\Gamma(n+1)\,\Gamma(1/2)}{\Gamma\left(\dfrac{n+3}{2}\right)}$$

$$= \frac{n\,!}{2^n} \frac{\Gamma\left(\dfrac{1}{2}\right)}{\left(n+\dfrac{1}{2}\right)\left(n-\dfrac{1}{2}\right)\left(n-\dfrac{3}{2}\right)\cdots\dfrac{1}{2}\,\Gamma\left(\dfrac{1}{2}\right)}$$

$$= \frac{2\,.\,n\,!\,2^n\,.\,n\,!}{(2n+1)\,!} = 2 \cdot \frac{2^n\,(n\,!)^2}{(2n+1)\,!},$$

i.e.,
$$I_n = \frac{2^{n+1}\,.\,(n\,!)^2}{(2n+1)\,!}. \tag{5.22}$$

If we now use the results (5.21) and (5.22) in succession, along with the fact that $P_m(x)$ is a polynomial of degree 'm', with the coefficient of the highest power of x i.e., x^m as (see equation (5.1)) $\dfrac{(2m)\,!}{2^m\,.\,(m\,!)^2}$, we easily find that

$$\left.\begin{aligned} &\int_{-1}^{1} P_m(x)\,P_n(x)\,dx = 0, && \text{if } m < n \\ &\text{and} \\ &\int_{-1}^{1} P_m(x)\,P_n(x)\,dx = \frac{2}{2n+1}, && \text{if } m = n. \end{aligned}\right\} \tag{5.23}$$

Interchanging the roles of m and n we also obtain that

$$\int_{-1}^{1} P_m(x)\,P_n(x)\,dx = 0, \text{ if } m > n. \tag{5.24}$$

The results (5.23) and (5.24) ultimately prove the theorem.

Application

Using the above orthogonality property of Legendre Polynomials, we can easily express any positive integral power of x, in terms of Legendre Polynomials, for $x \in (-1, 1)$, and we find that if we write

$$x^m = \sum_{n=0}^{m} a_n P_n(x), \quad m = 0, 1, 2, \ldots \tag{5.25}$$

then the coefficients a_n are given by

$$\begin{aligned} a_n &= \frac{2n+1}{2} \int_{-1}^{1} x^m P_n(x)\, dx \\ &= \frac{2n+1}{2} I_{m,n}, \end{aligned} \tag{5.26}$$

where the values of $I_{m,n}$ are given by (5.21) and (5.22).

Note : Expressing the power function x^m for $x \in (-1, 1)$ has enormous amount of advantages in the theory of partial differential equations associated with boundary value problems (see Williams [16]).

We give below some particular results for the purpose of ready visualisation and their derivations are left to the reader. We have that

$$\left\{ \begin{aligned} & x^0 = P_0(x), \; x = P_1(x), \; x^2 = \frac{1}{3}(2P_2(x) + P_0(x)), \\ & x^3 = \frac{1}{5}(2P_3(x) + 3P_1(x)), \; x^4 = \frac{1}{35}(8P_4(x) + 20P_2(x) + 7P_0(x)), \\ & x^5 = \frac{1}{63}(8P_5(x) + 28P_3(x) + 27P_1(x)), \\ & x^6 = \frac{16}{231}P_6(x) + \frac{24}{77}P_4(x) + \frac{10}{21}P_2(x) + \frac{1}{7}P_0(x), \\ & \text{etc.} \end{aligned} \right. \tag{5.27}$$

Using definition 5.1.1, we also find that the first few Legendre Polynomials are given by

$$\left\{ \begin{aligned} & P_0(x) = 1, \; P_1(x) = x, \; P_2(x) = \frac{1}{2}(3x^2 - 1), \\ & P_3(x) = \frac{1}{2}(5x^3 - 3x), \; P_4(x) = \frac{1}{8}(35x^4 - 30x^2 + 3), \\ & P_5(x) = \frac{1}{8}(63x^5 - 70x^3 + 15x), \\ & P_6(x) = \frac{1}{16}(231x^6 - 315x^4 + 105x^2 - 5), \\ & \text{etc.} \end{aligned} \right. \tag{5.28}$$

(B) An Integral Representation

We prove the following result (see Lebedev [8]).

Theorem 5.1.5

For all real values of x we have that

$$P_n(x) = \frac{1}{2\pi i}\int_c \frac{(u^2-1)^n}{2^n(u-x)^{n+1}}\,du, \tag{5.29}$$

where c is a closed contour in the complex u-plane, surrounding the point $u = x$.

Proof

We start with the function $w(x, t) = (1 - 2xt + t^2)^{-1/2}$, for complex values of the variable t, and for a fixed real value of x.

Using Taylor's expansion formula for a complex function we find that we can write

$$w(x, t) = \sum_{n=0}^{\infty} c_n(x)t^n, \; |t| < r, \tag{5.30}$$

where $r = \min(|r_1|, |r_2|)$, r_1, r_2 denoting the two roots of the quadratic equation $1 - 2xt + t^2 = 0$.

The Taylor coefficients $c_n(x)$ (see Churchill [2]),are given by

$$c_n(x) = \frac{1}{2\pi i}\int_L (1-2xt+t^2)^{-1/2}\, t^{-n-1}\,dt \tag{5.31}$$

where L is a closed contour surrounding the point $t = 0$ and lying inside the circular region $|t| < r$ of the complex t-plane.

We make the substitution

$$1 - ut = (1 - 2xt + t^2)^{1/2} \tag{5.32}$$

i.e.

$$(1-ut)^2 = 1 - 2xt + t^2$$

or

$$t^2(u^2-1) + 2t(x-u) = 0,$$

giving

$$t = 2(u-x)/(u^2-1), \tag{5.33}$$

and then

$$dt = 2\cdot\frac{(2ux-u^2-1)}{(u^2-1)^2}\,du$$

or

$$dt = \frac{2(1-ut)}{u^2-1}\,du, \text{ by using (5.33).} \tag{5.34}$$

Using (5.33) and (5.34), we finally obtain, from (5.31), the result that

$$c_n(x) = \frac{1}{2\pi i}\int_c \frac{(u^2-1)^n}{2^n(u-x)^{n+1}}\,du, \tag{5.35}$$

where c is the transformation of the contour L, and encloses the point $u = x$ (see (5.33)).

Evaluating the integral on the right hand-side of (5.35) by using Cauchy's residue theorem (see [2]) (note that $u = x$ is a pole of order $n + 1$), we ultimately prove that

$$c_n(x) = \frac{1}{2^n n!}\left[\frac{d^n}{du^n}\{u^2 - 1)^n\}\right]_{u=x} \equiv P_n(x),$$

and this is the desired result.

Application

1. Choosing the path c of *Theorem* 5.1.5 to be a circle of radius $\sqrt{|x^2 - 1|}$ with centre at the point $u = x$, i.e., choosing

$$u = x + \sqrt{|x^2 - 1|}\, e^{i\phi},\ -\pi \le \phi \le \pi,$$

we easily deduce that,

$$P_n(x) = \frac{1}{2\pi}\int_{-\pi}^{\pi}\left[\frac{x^2 + 2x\sqrt{|x^2 - 1|}\, e^{i\phi} + |x^2 - 1|\, e^{2i\phi} - 1}{2\sqrt{|x^2 - 1|}\, e^{i\phi}}\right]^n d\phi,$$

or

$$P_n(x) = \frac{1}{\pi}\int_0^{\pi}[x + \sqrt{(x^2 - 1)}\cos\phi]^n\, d\phi, \text{ if } |x| \ge 1. \tag{5.36}$$

This formula is called 'Laplace's integral' for Legendre polynomials.

It can also be proved that the formula (5.36) for $P_n(x)$ holds even if $|x| < 1$. We find that for $|x| \le 1$ we have

$$|x + \sqrt{x^2 - 1}\cos\phi| = |x + i\sqrt{1 - x^2}\cos\phi|$$
$$= \sqrt{x^2 + (1 - x^2)\cos^2\phi} \le 1,$$

and hence, we obtain an important result that

$$|P_n(x)| \le 1, \text{ for } -1 \le x \le 1. \tag{5.37}$$

2. For $-1 < x < 1$, we get $x = \cos\theta$, $(0 < \theta < \pi)$, in (5.36) and obtain

$$P_n(\cos\theta) = \frac{1}{\pi}\int_0^{\pi}(\cos\theta + i\sin\theta\cos\phi)^n\, d\phi. \tag{5.38}$$

Introducing a new complex variable of integration $\zeta = \cos\theta + i\sin\theta\cos\phi$, we derive that

$$P_n(\cos\theta) = \frac{1}{\pi i}\int_{e^{i\theta}}^{e^{-i\theta}}\frac{\zeta^n\, d\zeta}{\sqrt{1 - 2\zeta\cos\theta + \zeta^2}}, \tag{5.39}$$

where the integral is evaluated along the line-segment joining the points $\zeta = e^{-i\theta}$ and $\zeta = e^{+i\theta}$, both lying on the unit circle in the complex ζ-plane.

If we next use Cauchy's integral theorem involving shifting of contours of integration, we find that we can replace the formula (5.39), after writing $\zeta = e^{i\psi}$, by

$$P_n(\cos\theta) = \frac{1}{\pi}\int_{-\theta}^{\theta} \frac{e^{i(n+1/2)\psi}}{\sqrt{2\cos\psi - 2\cos\theta}}\,d\psi,$$

i.e.

$$P_n(\cos\theta) = \frac{\sqrt{2}}{\pi}\int_0^{\theta} \frac{\cos(n+1/2)\psi}{\sqrt{\cos\psi - \cos\theta}}\,d\psi. \tag{5.40}$$

Changing ψ to $\pi - \psi$ we also obtain that

$$P_n(\cos\theta) = \frac{\sqrt{2}}{\pi}\int_0^{\pi} \frac{\sin(n+1/2)\psi}{\sqrt{\cos\theta - \cos\psi}}\,d\psi. \tag{5.41}$$

The formulae (5.40) and (5.41) are known as the 'Mehler-Dirichlet' formulae for Legendre polynomials. There are several important applications of the above formulae in boundary value problems of Mathematical Physics (see Williams [16] and Sneddon [14], for example).

5.2 LEGENDRE FUNCTION OF THE SECOND KIND

We shall be concerned in this book only with Legendre functions of non-negative integral order, and we shall define:

Definition 5.2.1

The function

$$Q_n(x) = \frac{1}{2}\int_{-1}^{1} \frac{P_n(y)}{x-y}\,dy, \quad (|x| \neq 1) \tag{5.42}$$

is defined to be the Legendre function of the 'second kind' and of non-negative integral order 'n' where $P_n(x)$ is the Legendre function of the 'first kind' and of the same order n.

We easily obtained from (5.42) that, for $n \geq 1$.

$$\begin{aligned} Q_n(x) &= \frac{1}{2}\int_{-1}^{1} \frac{P_n(x)}{x-y}\,dy + \frac{1}{2}\int_{-1}^{1} \frac{P_n(y) - P_n(x)}{x-y}\,dy \\ &= \frac{1}{2} P_n(x) \log\left|\frac{x+1}{x-1}\right| - W_{n-1}(x) \quad \text{(say)} \end{aligned} \tag{5.43}$$

where $W_{n-1}(x)$ is a polynomial of degree $n - 1$ $(n \geq 1)$.

Note that for $n = 0$, we have that

$$Q_0(x) = \frac{1}{2}\int_{-1}^{1} \frac{dy}{x-y} = \frac{1}{2}\log\left|\frac{x+1}{x-1}\right| \equiv \frac{1}{2} P_0(x) \log\left|\frac{x+1}{x-1}\right|. \tag{5.44}$$

Before proceeding further, we quote below the explicit forms of the first few values of $Q_n(x)$ obtained by a straightforward use of the expressions (5.28) for the functions $P_n(x)$:

$$\left.\begin{aligned} Q_0(x) &= \frac{1}{2}\log\left|\frac{x+1}{x-1}\right|, \\ Q_1(x) &= xQ_0(x) - 1 = P_1(x)\,Q_0(x) - P_0(x) \\ Q_2(x) &= P_2(x)\,Q_0(x) - \frac{3}{2}x = P_2(x)\,Q_0(x) - \frac{3}{2}P_1(x) \\ Q_3(x) &= P_3(x)\,Q_0(x) - \frac{5}{3}P_1(x) - \frac{1}{6}P_0(x) \\ Q_4(x) &= P_4(x)\,Q_0(x) - \frac{7}{4}P_3(x) - \frac{1}{3}P_1(x), \quad \text{etc.} \end{aligned}\right\} \tag{5.45}$$

We next prove the following basic recurrence relations for the functions $Q_n(x)$.

Theorem 5.2.1

The Legendre functions of the second kind, $Q_n(x)$, satisfy the recurrence relations

(i) $Q'_{n+1}(x) - xQ'_n(x) = (n+1)\,Q_n(x)$

and

(ii) $xQ'_n(x) - Q'_{n-1}(x) = n\,Q_n(x)$,

for $n = 1, 2, 3, \ldots$

Proof

The definition 5.2.1 alongwith the identity

$$\frac{d}{dy}\log|y-x| = \frac{1}{y-x}$$

immediately give, after an integration by parts, that

$$Q_n(x) = -\frac{1}{2}[P_n(1)\log|1-x| - P_n(-1)\log|1+x| - \int_{-1}^{1} P'_n(y)\log|y-x|\,dy].$$

Differentiating this result with respect to x and using the results that

$$P_n(1) = 1,\; P_n(-1) = (-1)^n,$$

we obtain that

$$Q'_n(x) = -\frac{1}{2}\left[\frac{1}{x-1} - \frac{(-1)^n}{x+1} - \int_{-1}^{1}\frac{P'_n(y)\,dy}{x-y}\right]. \tag{5.46}$$

We thus find that

$$Q'_{n+1}(x) - Q'_{n-1}(x) = \frac{1}{2}\int_{-1}^{1}\frac{P'_{n+1}(y) - P'_{n-1}(y)}{x-y}\,dy,$$

and this, in turn, gives by using the recurrence relation (iv) for $P_n(x)$, deduced earlier, that

$$Q'_{n+1}(x) - Q'_{n-1}(x) = (2n+1)\, Q_n(x). \tag{5.47}$$

Now, from the above few results we obtain

$$\begin{aligned} nQ_n(x) + Q'_{n-1}(x) &= \frac{n}{2}\int_{-1}^{1} \frac{P_n(y)\,dy}{x-y} - \frac{1}{2}\left[\frac{1}{x-1} - \frac{(-1)^{n-1}}{x+1} - \int_{-1}^{1} \frac{P'_{n-1}(y)\,dy}{x-y}\right] \\ &= \frac{1}{2}\int_{-1}^{1} \frac{nP_n(y) + P'_{n-1}(y)}{x-y}\,dy - \frac{1}{2}\left[\frac{1}{x-1} - \frac{(-1)^{n-1}}{x+1}\right]. \end{aligned} \tag{5.48}$$

But we have deduced earlier a recurrence relation for $P_n(x)$ which is given by

$$nP_n(x) + P'_{n-1}(x) = xP'_n(x). \tag{5.49}$$

Using (5.49) in (5.48) we easily obtain

$$\begin{aligned} nQ_n(x) + Q'_{n-1}(x) &= x\left[\frac{1}{2}\int_{-1}^{1} \frac{P'_n(y)}{x-y}\,dy\right] \\ &\quad - \frac{1}{2}\left[\left(1 + \frac{1}{x-1}\right) - (-1)^n\left(1 - \frac{1}{x+1}\right)\right] \\ &= \frac{x}{1}\left[\int_{-1}^{1} \frac{P'_n(y)}{x-y}\,dy - \left\{\frac{1}{x-1} - \frac{(-1)^n}{x+1}\right\}\right] \\ &= xQ'_n(x), \text{ by using (5.46).} \end{aligned} \tag{5.50}$$

This proves the recurrence relation (ii). The relation (i) can be proved by eliminating $Q'_{n-1}(x)$ between the results (5.47) and (5.50).

As in the case of $P_n(x)$, we can easily prove that the function $Q_n(x)$ satisfies the Legendre differential equation.

$$\frac{d}{dx}\left[(1-x^2)\frac{dQ_n}{dx}\right] + n(n+1)\,Q_n = 0, \tag{5.51}$$

by just using the two recurrence relations (i) and (ii) derived above,with the exception of the two points $x = +1$ and $x = -1$, at which the function $Q_n(x)$ ceases to be defined.

Combining the two results (5.15) and (5.51) we thus observe that the following theorem holds:

Theorem 5.2.2

For all non-negative integers n and for $x \neq \pm 1$, the two Legendre functions $P_n(x)$ and $Q_n(x)$ represent the two linearly independent solutions of the Legendre differential equation

$$(1-x^2)\frac{d^2 y}{dx^2} - 2x\frac{dy}{dx} + n(n+1)\,y = 0.$$

We shall next obtain an explicit expression for the polynomial $W_{n-1}(x)$ occurring in equation (5.43). We prove that

Theorem 5.2.3

If, positive integral values of n,

$$Q_n(x) = \frac{1}{2}P_n(x)\log\left|\frac{x+1}{x-1}\right| - W_{n-1}(x),$$

then

$$W_{n-1}(x) = \sum_{s=0}^{m}\frac{(2n-4s-1)}{(2s+1)(n-s)}P_{n-2s-1}(x),$$

with $m = \frac{n}{2} - 1$ or $\frac{n-1}{2}$, according as n is even or odd.

Proof

We write

$$W_{n-1}(x) = P_n(x)\,Q_0(x) + Q_n(x).$$

Then using the fact that $P_n(x)$ and $Q_n(x)$ satisfy the Legendre's equation for $n = 0, 1, 2, \ldots.$ we easily show that

$$\left[(1-x^2)\frac{d^2}{dx^2} - 2x\frac{d}{dx} + n(n+1)\right]W_{n-1}(x) = 2P'_n(x). \qquad (5.52)$$

We next write the following two different expansions for the two polynomials $W_{n-1}(x)$ and $P'_n(x)$, of degree $n-1$:

$$W_{n-1}(x) = \sum_{r=0}^{n-1} c_r\,P_r(x) \qquad (5.53)$$

and

$$P'_n(x) = \sum_{s=0}^{n-1} a_s\,P_s(x), \qquad (5.54)$$

where the coefficients a_s can be determined by using the orthogonality relation of Theorem 5.1.4 and the result (5.21) as given by

$$a_s = \frac{2s+1}{2}\int_{-1}^{1} P'_n(x)\,P_s(x)\,dx$$

$$= \frac{2s+1}{2}\left[\{P_n(x)\,P_s(x)\}_{x=-1}^{1} - \int_{-1}^{1} P_n(x)\,P'_s(x)\,dx\right]$$

$$= \frac{2s+1}{2}\,[1-(-1)^{n-s}]. \qquad (5.55)$$

Using the results (5.53) to (5.55) in (5.52) along with the fact that $P_r(x)$ satisfies the Legendre equation with n replaced by r, and identifying the coefficients of $P_r(x)$ from both sides of the resulting equation, the theorem follows:

We complete the discussion on the two Legendre functions introduced before by noting that if n is a negative integer in the Legendre differential equation

$$\frac{d}{dx}\left[(1-x^2)\frac{dy}{dx}\right] + n(n+1)\,y = 0, \tag{5.56}$$

then setting $n = -(p+1)$, $p \geq 0$, we can rewrite the equation (5.56) as

$$\frac{d}{dx}\left[(1-x^2)\frac{dy}{dx}\right] + p(p+1)\,y = 0, \tag{5.57}$$

and the two linearly independent solutions of this last equation can be taken to be $P_p(x)$ and $Q_p(x)$ respectively, except when $x = \pm 1$.

For other values of the parameter n in the Legendre equation (5.56), we can determine its solution by employing the Frobenius' method discussed in Chapter 4.

We next take up the study of the "Associated Legendre equation" very briefly.

5.3 THE ASSOCIATED LEGENDRE FUNCTIONS

Definition 5.3.1

The ordinary differential equation

$$(1-x^2)\frac{d^2y}{dx^2} - 2x\frac{dy}{dx} + \left\{n(n+1) - \frac{m^2}{1-x^2}\right\}y = 0, \tag{5.58}$$

where n and m are known constants, is known as the "Associated Legendre equation".

We easily establish the following theorem :

Theorem 5.3.1

When m is a positive integer and n is a non-negative integer, then the two functions

$$\left.\begin{aligned} P_n^m(x) &= (-1)^m (1-x^2)^{m/2}\frac{d^m P_n(x)}{dx^m} \\ \text{and} \qquad & \\ Q_n^m(x) &= (-1)^m (1-x^2)^{m/2}\frac{d^m Q_n(x)}{dx^n} \end{aligned}\right\} \tag{5.59}$$

can be taken to be two linearly independent solutions of the Associated Legendre equation.

Proof

Differentiating the Legendre equation (5.56) m times, we get,

$$\left[(1-x^2)\frac{d^2}{dx^2} - 2(1+m)\frac{d}{dx} + (n-m)(n+m+1)\right]\frac{d^m y}{dx^m} = 0. \quad (5.60)$$

Also setting $y = (1-x^2)^{m/2}(-1)^m z$ in the associated Legendre equation (5.58) we arrive at the equation

$$\left[(1-x^2)\frac{d}{dx^2} - 2(1+m)\frac{d}{dx} + (n-m)(n+m+1)\right] z = 0. \quad (5.61)$$

Identifying equation (5.61) with equation (5.60) and using the fact that $P_n(x)$ and $Q_n(x)$ are two linearly independent solutions of the Legendre equation (5.56), the theorem follows.

We shall next prove the following orthogonality relation for the associated Legendre functions of the first kind.

Theorem 5.3.2

The associated Legendre functions $P_n^m(x)$ satisfy the orthogonality relation

$$\int_{-1}^{1} P_n^m(x)\, P_r^m(x)\, dx = \frac{2\,\Gamma(n+m+1)}{(2n+1)\,\Gamma(n-m+1)}\,\delta_{r,n},$$

for a fixed value of the non-negative integer m, where $\delta_{r,n}$, is the Krönecker delta.

Proof

We use the fact that $P_n^m(x)$ and $P_r^m(x)$ satisfy the associated Legendre equation, i.e.,

$$\left[(1-x^2)\frac{d^2}{dx^2} - 2x\frac{d}{dx} + \left\{n(n+1) - \frac{m^2}{1-x^2}\right\}\right] P_n^m(x) = 0 \quad (5.62)$$

and

$$\left[(1-x^2)\frac{d^2}{dx^2} - 2x\frac{d}{dx} + \left\{r(r+1) - \frac{m^2}{1-x^2}\right\}\right] P_r^m(x) = 0. \quad (5.63)$$

Multiplying equation (5.62) by $P_r^m(x)$ and equation (5.63) by $P_n^m(x)$ and taking difference we easily find that

$$\frac{d}{dx}\left[(1-x^2)\left\{P_r^m(x)\frac{dP_n^m(x)}{dx} - P_n^m(x)\frac{dP_r^m(x)}{dx}\right\}\right]$$

$$= -\{n(n+1) - r(r+1)\}P_n^m(x)\,P_r^m(x). \quad (5.64)$$

Integrating both sides of (5.64) with respect to x between the limits $x = -1$ and $x = +1$, we then obtain that

$$\int_{-1}^{1} P_n^m(x)\, P_r^m(x)\, dx = 0, \quad \text{if } n \neq r, \tag{5.65}$$

proving the first part of the theorem, for values of n not equal to r.

To prove the theorem for values of n equal to r, we proceed as follows:

We have, by using Rodrigue's formula (Theorem 5.1.1) that

$$\frac{d}{dx}[P_n^m(x)] = \frac{1}{2^n\, n!}\frac{d}{dx}\left[(1-x^2)^{m/2}\frac{d^{m+n}}{dx^{m+n}}\{(x^2-1)^n\}\right]$$

$$= \frac{1}{2^n n!}(1-x^2)^{m/2}\frac{d^{m+n+1}}{dx^{m+n+1}}\{(x^2-1)^n\}$$

$$-\frac{mx}{(1-x^2)}\frac{1}{2^n n!}(1-x^2)^{m/2}\frac{d^{m+n}}{dx^{m+n}}\{(x^2-1)^n\}$$

i.e.,

$$\frac{d}{dx}[P_n^m(x)] = \frac{1}{\sqrt{1-x^2}}P_n^{m+1}(x) - \frac{mx}{(1-x)^2}P_n^m(x);$$

therefore, we obtain the identity

$$\{P_n^{m+1}(x)\}^2 = (1-x^2)\left\{\frac{d}{dx}P_n^m(x)\right\}^2 + mx\frac{d}{dx}[\{P_n^m(x)\}^2]$$

$$+\frac{m^2x^2}{(1-x^2)}\{P_n^m(x)\}^2. \tag{5.66}$$

Thus, if we use the notation that

$$I_{n,m} = \int_{-1}^{1} P_n^m(x)\, P_n^m(x)\, dx, \tag{5.67}$$

we find, by using (5.66), that

$$I_{n,m+1} = \int_{-1}^{1}\left[(1-x^2)\left\{\frac{d}{dx}(P_n^m(x))\right\}\right]\left[\frac{d}{dx}(P_n^m(x))\right]dx$$

$$+ m\int_{-1}^{1} x\frac{d}{dx}\{(P_n^m(x))^2\}\,dx + m^2\int_{-1}^{1}\frac{x^2}{(1-x^2)}\{P_n^m(x)\}^2\,dx.$$

Integrating the first integral,in above, by parts we arrive at the result

$$I_{n,m+1} = -\int_{-1}^{1}\frac{d}{dx}\left[(1-x^2)\frac{d}{dx}(P_n^m(x))\right]P_n^m(x)\,dx$$

$$+ m\int_{-1}^{1} x\frac{d}{dx}\{(P_n^m(x))^2\}\,dx + m^2\int_{-1}^{1}\frac{x^2}{(1-x^2)}\{P_n^m(x)\}^2\,dx.$$

Using the associated Legendre equation (5.62), in the first integral above, and integrating the second integral by parts, we find that

$$I_{n,m+1} = -\int_{-1}^{1}\left[-\left\{n(n+1)-\frac{m^2}{1-x^2}\right\}(P_n^m(x))^2\right]dx$$

$$-m\int_{-1}^{1}\{(P_n^m(x))^2\}\,dx + m^2\int_{-1}^{1}\frac{x^2}{(1-x^2)}\{P_n^m(x)\}^2\,dx.$$

$$= [n(n+1)-m^2-m]\,I_{n,m}$$

i.e.,

$$I_{n,m+1} = (n-m)(n+m+1)\,I_{n,m}. \tag{5.68}$$

Equation (5.68) provides us with a recurrence relation for the integral $I_{n,m}$ we are interested in evaluating, and a repeated use of this recurrence relation gives (assuming that $m < n+1$) that

$$I_{n,m} = [(n-m+1)(n-m+2)\ldots n]\,[(n+m)(n+m-1)\ldots(n+1)]\,I_{n,0}$$

$$= \frac{n!}{(n-m)!}\,\frac{(n+m)!}{n!}\,I_{n,0} = \frac{\Gamma(n+m+1)}{\Gamma(n-m+1)}\,I_{n,0}. \tag{5.69}$$

But, we have (see Theorem 5.1.4) that

$$I_{n,0} \equiv \int_{-1}^{1} P_n(x)\,P_n(x)\,dx = \frac{2}{2n+1}. \tag{5.70}$$

Using (5.70) and (5.69), we thus have that

$$I_{n,m} \equiv \int_{-1}^{1} P_n^m(x)\,P_n^m(x)\,dx = \frac{2}{2n+1}\,\frac{\Gamma(n+m+1)}{\Gamma(n-m+1)}. \tag{5.71}$$

The two results (5.65) and (5.71) completely prove the theorem.

Note : The proof given above is based on the assumption that $m < n+1$. For value of $m \geq n+1$, we have, from the result (5.59) that $P_n^m(x)$ is identically equal to zero, and this is consistent with the result (5.71), if one uses the fact that $\Gamma(p) \to \infty$ if p takes on the values $p = 0, -1, -2, -3$ etc. Thus, the theorem holds for all values of the non-negative integers n, m and r.

PROBLEMS

1. Prove that the Legendre polynomials $P_n(x)$ and the Hermite polynomials $H_n(x)$ are connected by the relation:

$$P_n(x) = \frac{2}{\sqrt{\pi}\,n!}\int_0^\infty t^n e^{-t^2} H_n(xt)\,dt.$$

2. Show that any polynomial of degree n is a linear combination of the Legendre Polynomials $P_0, P_1 \ldots, P_n$.

3. Prove that

$$\frac{dP_n(x)}{dx} = (2n-1)\,P_{n-1}(x) + (2n-5)\,P_{n-3}(x) + (2n-9)\,P_{n-5}(x) + \ldots$$

4. Evaluate

$$\int_{-1}^{1} \log (1-x)\, P_n(x)\, dx.$$

5. Show that

$$\int_{-1}^{1} \frac{P_n^m(x)\, P_r^m(x)}{1-x^2}\, dx = \frac{\Gamma(n+m+1)}{m\,\Gamma(n-m+1)}\, \delta_{n,\,r}$$

6. Find the general solution of the ODE

$$\frac{d^2u}{d\theta^2} + \cot\theta \frac{du}{d\theta} + 2u = 0, \quad (0<\theta<\pi)$$

in terms of Legendre functions.

7. Show that if $f(\theta)$ is defined in $0 \le \theta \le \pi$, and absolutely integrable there, then

$$f(\theta) = \frac{1}{2} \sum_{n=0}^{\infty} (2n+1)\, P_n(\cos\theta) \int_0^{\pi} f(\theta)\, P_n(\cos\theta) \sin\theta\, d\theta.$$

$$\text{If} \quad f(\theta) = \begin{cases} 1, & (0<\theta<\alpha) \\ \theta, & (\alpha<\theta<\pi), \end{cases}$$

show that

$$f(\theta) = \frac{1}{2}(1-\cos\alpha) + \frac{1}{2} \sum_{n=1}^{\infty} \{P_{n-1}(\cos\alpha) - P_{n+1}(\cos\alpha)\}\, P_n(\cos\theta).$$

8. Prove that if $x > 1$,

$$Q_n(x) = \frac{1}{2^{n+1}} \int_{-1}^{1} \frac{(1-t^2)^n}{(x-t)^{n+1}}\, dt,$$

and deduce that

$$Q_n(x) = \int_0^{\alpha} \{x - (x^2-1)^{1/2} \cosh\theta\}^n\, d\theta,$$

where

$$\alpha = \frac{1}{2} \log (x+1)/(x-1).$$

9. Prove that the Wronskian $W[P_n(x), Q_n(x)]$ of the two solutions of Legendre's equation is given by

$$W[P_n(x), Q_n(x)] \equiv P_n(x)\, Q_n'(x) - P_n'(x)\, Q_n(x) = \frac{1}{1-x^2}.$$

Using this relation and the result that $Q_n(x) \to 0$ as $x \to \infty$ (follows from Definition 5.2.1). deduce that

$$Q_n(x) = P_n(x) \int_x^{\infty} \frac{d\xi}{(\xi^2 - 1)\{P_n(\xi)\}^2}.$$

10. Prove the following results: (for $|x| > 1$)

(i) $x^m Q_n(x) = \dfrac{1}{2} \displaystyle\int_{-1}^{1} \frac{t^m P_n(t)\,dt}{x - t}$,

where m, n are positive integers; with $m \leq n$, and

(ii) $x^{n+1} Q_n(x) = \dfrac{1}{2} \displaystyle\int_{-1}^{1} \frac{t^{n+1} P_n(t)}{x - t}\,dt + \frac{2^n (n!)^2}{(2n+1)!}$.

Deduce that

(a) if $m \leq n$,

$$P_m(x) Q_n(x) = \frac{1}{2} \int_{-1}^{1} \frac{P_m(t) P_n(t)}{x - t}\,dt.$$

(b) $P_{n+1}(x) Q_n(x) = \dfrac{1}{2} \displaystyle\int_{-1}^{1} \frac{P_{n+1}(t) P_n(t)}{x - t}\,dt + \frac{1}{n+1}$.

Replacing n by $n + 1$, m by n in (a), and subtracting the result from (b), show that

$$P_{n+1}(x) Q_n(x) - P_n(x) Q_{n+1}(x) = \frac{1}{n+1}.$$

11. Show that, if $|x| > 1$, $m > 1$, $n > -1$, then

$$Q_n^m(x) = \frac{\Gamma(n+m+1)}{\Gamma(n+1)} \frac{(x^2-1)^{1/2\,m}}{2^{n+1}} \int_{-1}^{1} \frac{(1-t^2)^n\,dt}{(x-t)^{n+m+1}}.$$

CHAPTER - SIX

The Laguerre Function

In this chapter we shall discuss, very briefly, some of the important properties of the solution of a special differential equation, known as the Laguerre differential equation.

We start with the following definition :

6.1 THE LAGUERRE POLYNOMIAL

Definition 6.1.1

The ordinary differential equation

$$x\frac{d^2y}{dx^2} + (1-x)\frac{dy}{dx} + \nu y = 0 \tag{6.1}$$

where ν is a constant (real or complex), is known as the Laguerre differential equation.

It is a straightforward matter to check that $x = 0$ is a regular singular point of equation (6.1) and that a direct application of Frobenius method produces the two linearly independent solutions of equation (6.1), as given by

$$y_1(x) = 1 + \frac{(-\nu)}{1!}\cdot\frac{x}{1!} + \frac{(-\nu)(-\nu+1)}{2!}\cdot\frac{x^2}{2!} + \frac{(-\nu)(-\nu+1)(-\nu+2)}{3!}\cdot\frac{x^3}{3!} + \ldots, \tag{6.2}$$

and

$$y_2(x) = y_1(x)\log x + \sum_{r=0}^{\infty} c_r x^r \tag{6.3}$$

with

$$c_r = \frac{(-\nu)(-\nu+1)\ldots(-\nu+r+1)}{(r!)^2}\left[\frac{1}{-\nu} + \frac{1}{1-\nu} + \ldots + \frac{1}{r-1-\nu}(-2)\left(1+\frac{1}{2}+\ldots+\frac{1}{r}\right)\right]. \tag{6.4}$$

We easily observe that, if the parameter ν of the Laguerre equation happens to be a non-negative integer n, then the solution $y_1(x)$ reduces to a polynomial of degree n, and we define:

Definition 6.1.2

The Laguerre polynomial $L_n(x)$ of degree n, is defined by the relation

$$L_n(x) = n!\,\{y_1(x)\}\ \nu = n \tag{6.5}$$

where $y_1(x)$ is given by the relation (6.2).

Using the symbol $(\alpha)_k$ to denote the expression

$$(\alpha)_k = \frac{\Gamma(\alpha + k)}{\Gamma(\alpha)}, \tag{6.6}$$

as in equation (5.7) before, we find that we can write

$$L_n(x) = n!\sum_{k=0}^{n} \frac{(-n)_k}{(k!)^2} x^k. \tag{6.7}$$

The first few Laguerre polynomials are quoted below for ready reference:

$$\begin{cases} L_0(x) = 1, \quad L_1(x) = 1 - x, \quad L_2(x) = 2 - 4x + x^2, \\ L_3(x) = 6 - 18x + 9x^2 - x^3, \\ L_4(x) = 24 - 96x + 72x^2 - 6x^3 + x^4, \quad \text{etc.} \end{cases} \tag{6.8}$$

As in the case of Legendre polynomials, discussed in Chapter 5, we have the following generating function representation of the Laguerre polynomials.

Theorem 6.1.1

The expansion

$$\frac{1}{(1-t)} \exp\left(-\frac{xt}{1-t}\right) = \sum_{n=0}^{\infty} \frac{t^n}{n!} L_n(x) \tag{6.9}$$

holds for all values of the real variable x and for values of t such that $0 < t < 1$.

Proof

Using the standard expansion formula for the exponential function we obtain

$$\frac{1}{(1-t)} \exp\left(\frac{-xt}{1-t}\right) = \sum_{r=0}^{\infty} \frac{(-1)^r x^r t^r}{r!\,(1-t)^{r+1}}$$

Using the Binomial expansion next we obtain

$$\frac{1}{(1-t)} \exp\left(\frac{-xt}{1-t}\right) = \sum_{r=0}^{\infty} \sum_{s=0}^{\infty} \frac{(-1)^r (r+1)_s}{r!\,s!} x^r t^{r+s} \tag{6.10}$$

with

$$(r+1)_s = (r+1)(r+2)\ldots(r+s) \tag{6.11}$$

when the relation (6.6) is also utilized.

The coefficient of t^n $(n \geq 0)$ of the expansion in the right of equation (6.10) is found to be given by

$$f_n(x) = \sum_{r=0}^{n} \frac{(-1)^r (r+1)_{n-r}}{r!\,(n-r)!} x^r, \tag{6.12}$$

so that (6.10) gives that

$$\frac{1}{(1-t)} \exp\left(-\frac{xt}{1-t}\right) = \sum_{n=0}^{\infty} f_n(x)\, t^n. \tag{6.13}$$

We shall next use the two results :

$$\text{(i)} \quad (r+1)_{n-r} = \frac{\Gamma(r+1+n-r)}{\Gamma(r+1)} = \frac{n!}{r!}, \tag{6.14}$$

and

$$\text{(ii)} \quad \frac{(-1)^r}{(n-r)!} = \frac{(-n)_n}{n!}, \tag{6.15}$$

and prove ultimately that the function $f_n(x)$ in equation (6.12) is given by

$$f_n(x) = \sum_{r=0}^{n} \frac{(-n)_r}{(r!)^2} x^r = \frac{L_n(x)}{n!}. \tag{6.16}$$

This, then proves the theorem completely.
We define :

Definition 6.1.3

The function $[1/(1-t)] \exp(-xt/1-t)$ is called the 'generating function' for the Laguerre polynomials.

We easily prove the following Rodrigue's type formula for the polynomials $L_n(x)$.

Theorem 6.1.2

The formula

$$L_n(x) = e^x \frac{d^n}{dx^n}(x^n e^{-x}) \tag{6.17}$$

holds.

Proof

Using Leibnitz's rule of differentiation of a product of two functions we obtain

$$e^x \frac{d^n}{dx^n}(x^n e^{-x}) = e^x \sum_{r=0}^{n} (-1)^r \frac{(-n)_r}{r!} (D^{n-r} x^n)(D^r e^{-x}). \tag{6.18}$$

$$\left[D \equiv \frac{d}{dx}\right].$$

But we have that

$$\left.\begin{aligned} &e^x D^r (e^{-x}) = (-1)^r, \\ \text{and} \quad &D^{n-x^r} = n!\,\frac{x^r}{r!}. \end{aligned}\right\} \tag{6.19}$$

Using (6.19) in (6.18), we finally arrive at the result (6.17) easily.

As in the cases of the Hermite and the Legendre polynomials, we can easily prove the following important recurrence relations, satisfied by the Laguerre polynomials, by using the theorems 6.1.1 and 6.1.2.

(a) $L_{n+1}(x) = (2n + 1 - x) L_n(x) - n^2 L_{n-1}(x)$

and

(b) $L'_n(x) = n[L'_{n-1}(x) - L_{n-1}(x)]$.

6.2 THE LAGUERRE FUNCTION

In this section, we define a special function, known as the Laguerre function.

Definition 6.2.1

The function $\phi_n(x)$, defined by

$$\phi_n(x) = \frac{1}{n!} e^{-x/2} L_n(x), \tag{6.20}$$

is known as the Laguerre function.

We have the following important orthogonality property satisfied by the set $\{\phi_n(x)\}$ of the Laguerre functions:

Theorem 6.2.1

The set $\{\phi_n(x)\}$ forms an orthonormal set in the semi-infinite interval $(0, \infty)$ i.e.,

$$I_{m,n} = \int_0^\infty \phi_m(x)\,\phi_n(x)\,dx = \delta_{m,n}, \tag{6.21}$$

with $\delta_{m,n}$ denoting the Krönecker delta.

Proof

We first observe that

$$\int_0^\infty e^{-x} x^m L_n(x)\,dx = \int_0^\infty x^m \frac{d^n}{dx^n}(x^n e^{-x})\,dx, \quad \text{by (6.17)},$$

$$= (-1)^m \cdot m! \int_0^\infty \frac{d^{n-m}}{dx^{n-m}}(x^n e^{-x})\,dx,$$

obtained by repeated integration by parts, m-times.

Thus, if $n > m$, we have that

$$\int_0^\infty e^{-x} x^m L_n(x)\,dx = 0, \quad (n > m) \tag{6.22}$$

Now, $L_m(x)$ being a polynomial of degree m, can be expressed in the form

$$L_m(x) = \sum_{r=0}^{\infty} c_r x^r \tag{6.23}$$

and utlizing the result (6.22) we easily find that

$$I_{m,n} \equiv \frac{1}{m!\,n!} \int_0^{\infty} e^{-x} L_m(x)\, L_n(x)\, dx = 0, \text{ if } m < n. \tag{6.24}$$

For $m > n$, we can interchange the roles of m and n, and find that

$$I_{m,n} = 0, \quad \text{if } m \neq n, \tag{6.25}$$

thus proving one part of result (6.21).

We next show that

$$I_{n,n} = \frac{1}{(n!)^2} \int_0^{\infty} e^{-x} [L_n(x)]^2\, dx$$

$$= \frac{1}{(n!)^2} \int_0^{\infty} \frac{d^n}{dx^n}(x^n e^{-x}) \left[n! \sum_{r=0}^{n} \frac{(-n)_r}{(r!)^2} x^r \right] dx, \text{ by (6.17)}$$

$$= \frac{1}{(n!)^2} \int_0^{\infty} \left[\frac{d^n}{dx^n}(x^n e^{-x}) \right] \frac{n!\,(-n)_n}{(n!)^2} x^n\, dx, \qquad \text{by (6.22)}$$

Using the identity (6.15), we ultimately find that

$$I_{n,n} = \frac{(-1)^n}{n!} \int_0^{\infty} x^n \frac{d^n}{dx^n}(x^n e^{-x})\, dx$$

$$= \frac{1}{n!} \int_0^{\infty} x^n e^{-x}\, dx, \qquad \text{(by repeated integration by parts, } n\text{-times).}$$

i.e.,

$$I_{n,n} = \frac{1}{n!} \Gamma(n+1) = 1. \tag{6.26}$$

The results (6.25) and (6.26) together prove the theorem.

We next define the special function, known as "Associated Laguerre polynomial". Its properties can be handled exactly in the similar manner as we have handled the "Associated Legendre functions".

Definition 6.2.2

The function

$$L_n^m(x) = \frac{d^m}{dx^m} L_n(x)$$

is known as the "Associated Laguerre function".

Differentiating the Laguerre differential equation

$$x \frac{d^2 y}{dx^2} + (1 - x) \frac{dy}{dx} + ny = 0 \tag{6.27}$$

m-times, with respect to x, we obtain the new differntial equation

$$x \frac{d^2 z}{dx^2} + (m + 1 - x) \frac{dz}{dx} + (n - m) z = 0, \tag{6.27}$$

with

$$z = \frac{d^m y}{dx^m}. \tag{6.28}$$

It follows directly from above that the "Associated Laguerre polynomial" $L_n^m(x)$ is a solution of the differential equation (6.27), which is known as the *Associated Laguerre differential equation*, and its study can be made in lines similar to those of the *Associated Legendre* equation.

PROBLEMS

1. (a) Prove that the functions

$$L_n^{(\alpha)}(x) = e^x \frac{x^{-\alpha}}{n!} \frac{d^n}{dx^n} (e^{-x} x^{n+\alpha}), \; n = 0, 1, 2, \ldots,$$

for arbitrary real $\alpha > -1$, define polynomials of degree n, and verify that

$$L_0^{(\alpha)}(x) = 1, \; L_1^{(\alpha)}(x) = 1 + \alpha - x,$$

$$L_2^{(\alpha)}(x) = \frac{1}{2} [(1 + \alpha)(2 + \alpha) - 2(2 + \alpha) x + x^2] \quad \text{etc.}$$

with the general result that

$$L_n^{(\alpha)}(x) = \sum_{k=0}^{n} \frac{\Gamma(n + \alpha + 1)}{\Gamma(k + \alpha + 1)} \cdot \frac{(-x)^k}{k!(n-k)!}.$$

Deduce that $L_n^{(0)}(x) = L_n(x)$.

Note : Levedev [8] uses the form $L_n^{(\alpha)}(x)$ as the definition of the Laguerre polynomials.

1. (b) Prove that

$$(1 - t)^{-\alpha = 1} e^{-xt/(1-t)} = \sum_{n=0}^{\infty} L_n^{(\alpha)}(x) t^n, \; |t| < 1.$$

Deduce the following recurrence relations:

(i) $L_n^{(\alpha+1)}(x) - L_{n-1}^{(\alpha-1)}(x) = L_n^{(\alpha)}(x), \; n = 1, 2, \ldots$

(ii) $x \dfrac{dL_n^{(\alpha)}(x)}{dx} = nL_n^{(\alpha)}(x) - (n + \alpha) L_{n-1}^{(\alpha)}(x), \; n = 1, 2, \ldots$

(iii) $\dfrac{dL_n^{(\alpha)}(x)}{dx} = -L_{n-1}^{(\alpha+1)}(x),\ n = 1, 2, \ldots,$

and show that $u = L_n^{(\alpha)}(x)$ solves the differential equation

$$xu''(x) + (\alpha + 1 - x)\, u'(x) + nu(x) = 0.$$

2. Show that the following two differential equations

(i) $xu''(x) + (\alpha + 1 - 2\nu)\, u'(x) + \left[n + \dfrac{\alpha+1}{2} - \dfrac{x}{4} + \dfrac{\nu(\nu-\alpha)}{x}\right] . u(x) = 0$

and

(ii) $u''(x) + \left[4n + 2\alpha + 2 - x^2 + \dfrac{\frac{1}{4} - \alpha^2}{x^2}\right] u(x) = 0$

have the particular solutions

(i) $u = e^{-x/2} . x^{\nu} L_n^{(\alpha)}(x)$

and

(ii) $u = e^{-x^2/2}\, x^{\alpha + 1/2}\, L_n^{(\alpha)}(x^2)$, respectively

3. Using the Orthonormality relations for the Laguerre functions $\{\phi_n(x)\}$, expand the following functions $f(x)$, in a series of $\phi_n(x)$.

(i) $f(x) = x^{\nu},\ \nu > -\dfrac{1}{2},$

(ii) $f(x) = e^{-ax},\ a > -\dfrac{1}{2},$

(iii) $f(x) = \displaystyle\int_0^{\infty} e^{-ax} (a+1)^{\alpha-1}\, da,\ \alpha > -\dfrac{1}{2},$

4. Derive the representation:

$$e^{-x/2} L_n(x) = \frac{1}{2^{n-1}\, n!\, \sqrt{\pi}} \int_0^{\infty} e^{-t^2} H_n^2(t) \cos(\sqrt{2xt})\, dt,$$

where $H_n(t)$ is the Hermite's polynomial of degree n.

CHAPTER - SEVEN

The Bessel Functions

The Bessel functions, like the other special functions of Mathematical Physics, i.e., the Hermite, the Legendre and the Laguerre Polynomials and functions, discussed in the previous chapters, have been extensively used in various branches of Applied Mathematics, and we shall only discuss here, in the present chapter some of the very basic properties of the various Bessel functions in a rather brief manner. We start with the very definition of the Bessel equations, from whose solutions these special functions have originated.

7.1 THE BESSEL EQUATION

Definition 7.1.1

The ordinary differential equation of the second order, as given by,

$$x^2 \frac{d^2 y}{dx^2} + x \frac{dy}{dx} + (x^2 - n^2)\, y = 0, \tag{7.1}$$

in which n is a constant, real or complex, is called the 'Bessel equation'.

It is immediately checked that the point $x = 0$ is a regular singular point of equation (7.1), and, therefore, Frobenius' method of obtaining its series solution, valid in a neighbourhood of $x = 0$, is applicable.

We assume, as in Chapter 4, that a series solution of equation (7.1) can be obtained in the form:

$$y(x, s) = \sum_{k=0}^{\infty} a_k x^{k+s}, \tag{7.2}$$

where the index s and the constants a_k are to be determined.

We obtain the 'indicial equation' in the form

$$s^2 - n^2 = 0, \tag{7.3}$$

giving the two roots

$$s_1 = -n \text{ and } s_2 = +n, \tag{7.4}$$

and then, various possible values of the constant n can be handled by employing the various ideas involved in the Frobenius' method.

Substituting (7.2) into (7.1) and equating the coefficients of x^{s+i} $(i \geq 1)$ to zero we obtain:

$$\left.\begin{aligned} & x^{s+1} : [(s+1)^2 - n^2]\, a_1 = 0, \\ \text{giving} \quad & a_1 = 0, \quad \text{for } s = \pm n, \text{ if } 2n \neq \pm 1, \end{aligned}\right\} \tag{7.5}$$

$$\left.\begin{aligned} & x^{s+2} : [(s+2)^2 - n^2]\, a_2 + a_0 = 0 \\ \text{giving} \quad & a_2 = -\frac{a_0}{(s+2+n)(s+2-n)} \\ & \text{(well defined for } s = \pm n, \text{ if } 2n \neq \pm 2). \end{aligned}\right\} \tag{7.6}$$

$$\left.\begin{aligned} & x^{s+3} : [(s+3)^2 - n^2]\, a_3 + a_1 = 0, \\ \text{giving} \quad & a_3 = -\frac{a_1}{(s+3+n)(s+3-n)} = 0 \\ \text{for} \quad & s = \pm n, \text{ if } 2n \neq \pm 3 \end{aligned}\right\} \tag{7.7}$$

and so on.

Thus, if $2n$ *is not an interger*, positive or negative, we find that the coeffficients a_k in (7.2) are obtained, for the root $s = +n$ of the indicial equation (7.3), in the form:

$$a_1 = a_3 = a_5 = \ldots = 0, \tag{7.8}$$

and

$$\left.\begin{aligned} a_2 &= (-1)^1 \frac{\Gamma(n+1)}{\Gamma(n+2)} \frac{1}{1!} \frac{1}{2^2}\, a_0 \\ a_4 &= (-1)^2 \frac{\Gamma(n+1)}{\Gamma(n+3)} \frac{1}{2!} \frac{1}{2^4}\, a_0 \\ a_6 &= (-1)^3 \frac{\Gamma(n+1)}{\Gamma(n+4)} \frac{1}{3!} \frac{1}{2^6}\, a_0 \\ &\cdots\cdots\cdots\cdots \\ a_{2r} &= (-1)^r \frac{\Gamma(n+1)}{\Gamma(n+r+1)} \frac{1}{r!} \frac{1}{2^{2r}}\, a_0. \end{aligned}\right\} \tag{7.9}$$

We, therefore, obtain one solution of the Bessel equation (7.1), as given by

$$y_1(x) = [y(x, s)]_{s = s_2 = n}$$

$$= a_0 \sum_{r=0}^{\infty} \frac{(-1)^r}{r!} \frac{\Gamma(n+1)\, x^n}{\Gamma(n+r+1)} (x/2)^{2r} \tag{7.10}$$

Note : In the case when $2n$ is a positive integer, we observe that $s_2 = n$ represents the greater of the two roots of the indicial equation (7.3), and the Frobenius' method suggests that the function $[y(x, s)]_{s = s_2 = n}$, *i.e.*, the expression on the right of (7.10) represents a valid solution of the Bessel equation (7.1). A similar argument can be employed even if $2n$ is a negative integer.

because then the greater root of the indicial equation turns out to be the root $s_1 = -n$, and the appropriate solution is given by $[y(x, s)]_{s = s_1 = -n}$.

7.1.1 The Bessel Functions

We define, in this section, the special functions, known as the Bessel functions, and derive certain basic properties of these functions.

Definition 7.1.2

The function $J_n(x)$, defined by the relation

$$J_n(x) = \sum_{r=0}^{\infty} \frac{(-1)^r}{\Gamma(r+1)\,\Gamma(n+r+1)} (x/2)^{2r+n} \tag{7.11}$$

is called the 'Bessel function' of the 'first kind', and of order 'n' real or complex.

From what has been derived and argued before, it is clear that we have the following result:

Theorem 7.1.1

The Bessel function of the first kind, i.e., the function $J_n(x)$ is one of the two solutions of the Bessel equation (7.1), for all values of the constant n, real or complex. Also $J_{-n}(x)$ represents a solution of the Bessel equation, which may or may not be independent of the solution $J_n(x)$, always.

In fact, if n is an integer, we have the following relation holding good:

Theorem 7.1.2

In n is an integer,

$$J_{-n}(x) = (-1)^n\, J_n(x), \tag{7.12}$$

showing thereby, that $J_n(x)$ and $J_{-n}(x)$ do not provide with the two independent solutions of the Bessel equation.

Proof

Using definition 7.1.2, we have that

$$J_n(x) = \sum_{s=0}^{\infty} \frac{(-1)^r}{\Gamma(r+1)\,\Gamma(-n+r+1)} (x/2)^{2r-n}. \tag{7.13}$$

Setting $r - n = s$, and using the fact that $\Gamma(-m) \to \infty$, for $m = 0, 1, 2, \ldots$, we obtain

$$\begin{aligned} J_{-n}(x) &= (-1)^n \sum_{s=0}^{\infty} \frac{(-1)^s}{\Gamma(s+1)\,\Gamma(n+s+1)} (x/2)^{2s+n} \\ &= (-1)^n J_n(x). \qquad \text{[by (7.11)]}. \end{aligned}$$

We note, while passing, that the two functions $J_n(x)$ and $J_{-n}(x)$ do represent the two linearly independent solutions of the Bessel equation always, when n is not an integer, as is obvious from the two forms of these functions as given by (7.11) and (7.13).

For general values of n, we have the following theorem :

Theorem 7.1.3

The two linearly independent solutions of the Bessel equation (7.1) may be taken to be the two functions $y_1(x)$ and $y_2(x)$, as given by

$$y_1(x) = J_n(x),$$

and

$$y_2(x) = \lim_{\nu \to n} \frac{\cos(\nu\pi) J_\nu(x) - J_{-\nu}(x)}{\sin \nu\pi} = Y_n(x) \quad \text{say}$$

Proof

That $J_n(x)$ is one solution of the Bessel equation, has been proved earlier and what remains to be proved is that $Y_n(x)$ is a solution of the Bessel equation, which is independent of the solution $J_n(x)$.

We observe that if n is not an integer, the function $Y_n(x)$ is given by :

$$Y_n(x) = \cot(n\pi) J_n(x) - J_{-n}(x) \operatorname{cosec}(n\pi), \tag{7.14a}$$

and this certainly represents a solution of the Bessel equation, which is independent of $J_n(x)$. Thus we have to just bother about the case when n is an integer.

We find that, if n is an integer, by applying *L'Hospital's rule, we have*

$$Y_n(x) = \frac{1}{\pi} \lim_{\nu \to n} \left[\frac{\partial J_\nu(x)}{\partial \nu} - (-1)^n \frac{\partial J_{-\nu}(x)}{\partial \nu} \right] \tag{7.14b}$$

Now, the functions $J_\nu(x)$ and $J_{-\nu}(x)$ satisfy the Bessel equation

$$x^2 \frac{d^2 y}{dx^2} + x \frac{dy}{dx} + (x^2 - \nu^2) y = 0. \tag{7.15}$$

Differentiating equation (7.15) w.r.t. ν and writing $y = J_\nu(x)$ once and $y = J_{-\nu}(x)$, next, we get

$$x^2 \frac{d^2}{dx^2}\left(\frac{\partial J_\nu(x)}{\partial \nu}\right) + x \frac{d}{dx}\left(\frac{\partial J_\nu(x)}{\partial \nu}\right) + (x^2 - \nu^2) \frac{\partial J_\nu(x)}{\partial \nu} - 2\nu J_\nu(x) = 0, \tag{7.16}$$

and

$$x^2 \frac{d^2}{dx^2}\left(\frac{\partial J_{-\nu}(x)}{\partial \nu}\right) + x \frac{d}{dx}\left(\frac{\partial J_{-\nu}(x)}{\partial \nu}\right) + (x^2 - \nu^2) \frac{\partial J_{-\nu}(x)}{\partial \nu} - 2\nu J_{-\nu}(x) = 0. \tag{7.17}$$

Multiplying equation (7.17) by $(-1)^n$ and subtracting the result from equation (7.16), we then arrive at the relation:

$$x^2 \frac{d^2}{dx^2}\left[\frac{\partial J_\nu(x)}{\partial \nu} - (-1)^n \frac{\partial J_{-\nu}(x)}{\partial \nu}\right] + x \frac{d}{dx}\left[\frac{\partial J_\nu(x)}{\partial \nu} - (-1)^n \frac{\partial J_{-\nu}(x)}{\partial \nu}\right] + (x^2 - \nu^2)\left[\frac{\partial J_\nu(x)}{\partial \nu} - (-1)^n \frac{\partial J_{-\nu}(x)}{\partial \nu}\right] - 2\nu\,[J_\nu(x) - (-1)^n J_{-\nu}(x)] = 0. \tag{7.18}$$

If we then take limits of both sides of equation (7.18), as $\nu \to n$ (an integer), and use theorem 7.1.2, we find that the function $Y_n(x)$, as given by (7.14b) really solves the Bessel equation, and that this solution $Y_n(x)$ is linearly independent of the solution $J_n(x)$, when n is an integer, is obvious from the form of the function $J_n(x)$, as given by equation (7.11), as well as the relation (7.14b).

We define :

Definition 7.1.3

The function $Y_n(x)$, as given by the relations (7.14a) and (7.14b), is called the 'Bessel function' of the 'second kind', and of order 'n'.

What all has been said and proved above, can be summarised by means of the following theorem :

Theorem 7.1.4

The general solution of the Bessel equation

$$x^2 \frac{d^2 y}{dx^2} + x \frac{dy}{dx} + (x^2 - n^2) y = 0 \quad (n = a \text{ constant})$$

can be taken in the form

$$y = y_{GS} = c_1 J_n(x) + c_2 Y_n(x),$$

where c_1 and c_2 are two arbitrary constants.

Using the fact that the series on the right hand side of equation (7.11) converges uniformly in n, we can differentiate it term by term and obtain

$$\left.\frac{\partial J_\nu(x)}{\partial \nu}\right|_{\nu = n \text{ (a positive integer)}}$$

$$= \sum_{k=0}^{\infty} \frac{(-1)^k (x/2)^{n+2k}}{\Gamma(k+1)\,\Gamma(n+k+1)} \left[\log \frac{x}{2} - \psi(k+n+1)\right] \tag{7.19}$$

where

$$\psi(z) = \frac{\Gamma'(z)}{\Gamma(z)} \equiv \frac{d}{dz}(\log \Gamma(z)). \tag{7.20}$$

We also obtain, in a similar fashion, that

$$\frac{\partial J_{-\nu}(x)}{\partial \nu} = \sum_{k=0}^{\infty} \frac{(-1)^k (x/2)^{-\nu+2k}}{\Gamma(k+1)\,\Gamma(k-\nu+1)} \left[-\log \frac{x}{2} + \psi(k-\nu+1)\right] \tag{7.21}$$

Now, we observe that, for $k = 0, 1, 2, \ldots, n-1$, we have that

$$\Gamma(k-\nu+1) \to \infty, \text{ and } \psi(k-\nu+1) \to \infty \text{ as } \nu \to n,$$

a positive integer.

But, it can be shown, by using standard formulas for the gamma function (see Appendix), that

$$\lim_{\nu \to n \text{ (a positive integer)}} \frac{\psi(k-\nu+1)}{\Gamma(k-\nu+1)}$$

$$= \lim_{\nu \to n} \left[\Gamma(\nu - k) \sin \pi (\nu - k) \frac{\psi(\nu - k) + \pi \cot \pi (\nu - k)}{\pi} \right]$$

$$= (-1)^{n-k} (n-k-1)\,!, \quad k = 0, 1, ..., n-1. \tag{7.22}$$

Using the result (7.22) in (7.21), and introducing a new summation index $p = k - n$, we find that

$$\left. \frac{\partial J_{-\nu}(x)}{\partial \nu} \right|_{\nu = n \text{ (a positive integer)}} = (-1)^n \sum_{k=0}^{n-1} \frac{(n-k-1)\,!}{k\,!} (x/2)^{2k-n}$$

$$+ (-1)^n \sum_{p=0}^{\infty} \frac{(-1)^p}{(n+p)\,!\,p\,!} \left[-\log \frac{x}{2} + \psi(p+1) \right] (x/2)^{2p+n}. \tag{7.23}$$

Using the result (7.20) and (7.23) in succession, in the formula (7.14b) for $Y_n(x)$, we ultimately prove the following result, written in the form of a theorem:

Theorem 7.1.5

The expansion formula for the Bessel function $Y_n(x)$, of second kind, is given by

$$Y_n(x) = -\frac{1}{\pi} \sum_{k=0}^{n-1} \frac{(n-k-1)\,!}{k\,!} (x/2)^{2k-n}$$

$$+ \frac{1}{\pi} \sum_{k=0}^{\infty} \frac{(-1)^k (x/2)^{n+2k}}{k\,!\,(n+k)\,!} \left[2 \log \frac{x}{2} - \psi(k+1) - \psi(k+n+1) \right]$$

$$(n = 0, 1, 2, ...) \tag{7.24}$$

with the understanding that the first sum should be set equal to zero if $n = 0$.

If n is a negative integer, we can use the identity

$$Y_{-n}(x) = (-1)^n Y_n(x), \tag{7.25}$$

to be obtained by using definition (7.14b), along with the theorem 7.1.2, and obtain useful expansions for $Y_n(x)$, from the formula (7.24).

If we next use the following results (see Appendix)

$$\psi(1) = -\gamma, \psi(m+1) = -\gamma + 1 + \frac{1}{2} + ... + \frac{1}{m}, (m = 1, 2, ...) \tag{7.26}$$

with $\gamma = 0.57721566...$, the Euler's constant, we find that the function $Y_n(x)$, for $n = 1, 2, ...$, can be expressed in the form:

$$Y_n(x) = \frac{2}{\pi} J_n(x) \log \frac{x}{2} - \frac{1}{\pi} \sum_{k=0}^{n-1} \frac{(n-k-1)\,!}{k\,!} (x/2)^{2k-n}$$

$$- \frac{1}{\pi} \sum_{k=0}^{\infty} \frac{(-1)^k (x/2)^{n+2k}}{k\,!\,(n+k)\,!} [\psi(k+1) + \psi(k+n+1)] \tag{7.27}$$

with the understanding that the second term on the right should be taken to be equal to zero, if $n = 0$.

7.1.2 The Recurrence Relations for the Bessel Functions

Using the definitions 7.1.2 and 7.1.3, we can prove the following recurrence relations satisfied by the two Bessel functions $J_n(x)$ and $Y_n(x)$, for all possible choices of the complex constant n.

Theorem 7.1.6

The Bessel functions $c_n(x)$ $(= J_n(x)$ or $Y_n(x))$ satisfy the recurrence relations :

(i) $xc'_n(x) = nc_n(x) - xc_{n+1}(x)$,

(ii) $\frac{2n}{x} c_n(x) = c_{u-1}(x) + c_{n+1}(x)$,

(iii) $2c'_n(x) = c_{n-1}(x) - c_{n+1}(x)$.

(iv) $xc'_n(x) = xc_{n-1}(x) - nc_n(x)$,

where 'dash' denotes differentiation with respect to x.

Proof

We shall demonstrate the proofs of the above recurrence relations, only when $c_n(x) = J_n(x)$, and leave it, as an exercise to the reader to prove that results when $c_n(x) = y_n(x)$

Proof of (i) $(c_n(x) = J_n(x))$

Using definition 7.1.2, we obtain

$$J'_n(x) = \frac{1}{2} \sum_{r=0}^{\infty} \frac{(-1)^r}{\Gamma(r+1)\,\Gamma(n+r+1)} (2r+n)\,(x/2)^{2r+n-1}$$

$$\therefore \quad xJ'_n(x) = nJ_n(x) + 2 \sum_{r=0}^{\infty} \frac{(-1)^r \cdot r}{\Gamma(r+1)\,\Gamma(n+r+1)} (x/2)^{2r+n}$$

$$= nJ_n(x) + 2 \sum_{r=1}^{\infty} \frac{(-1)^s}{\Gamma(r)\,\Gamma(n+r+1)} (x/2)^{2r+n}$$

(since $\Gamma(0) \to \infty$)

$$= n\,J_n(x) - 2 \sum_{s=0}^{\infty} \frac{(-1)^s}{\Gamma(s+1)\,\Gamma(n+s+2)} (x/2)^{2s+n+2}$$

(setting $r = s + 1$)

i.e.,

$$xJ'_n(x) = nJ_n(x) - xJ_{n+1}(x), \qquad \text{(using (7.11))} \tag{7.28}$$

and this proves the result (i), for $(c_n(x) = J_n(x))$

Proof of (iv) $(c_n(x) = J_n(x))$

Using (7.11), we obtain that

$$2\,J_n'(x) = \sum_{r=0}^{\infty} \frac{(-1)^r}{\Gamma(r+1)\,\Gamma(n+r+r)} \{2(n+r)-n\}\,(x/2)^{2r+n-1}$$

$$= 2\sum_{r=0}^{\infty} \frac{(-1)^r}{\Gamma(r+1)\,\Gamma(n+r)} (x/2)^{2r+n-1} - n\cdot\frac{2}{x} J_n(x).$$

$$= 2\,J_{n-1}(x) - \frac{2n}{x} J_n(x).$$

i.e., $$x\,J'_n(x) = x\,J_{n-1}(x) - nJ_n(x), \tag{7.29}$$

which proves the result (iv), if $c_n(x) = J_n(x)$.

Proofs of (ii) and (iii) ($c_n(x) = J_n(x)$).

Eliminating $xJ'_n(x)$ between (7.28) and (7.29) we can easily prove the result (ii). The result (iii) similarly can be proved by eliminating $nJ_n(x)$ between (7.28) and (7.29).

A straightforward use of the above recurrence relations for the Bessel functions $J_n(x)$ is given by the following theorem.

Theorem 7.1.7

If n is a positive integer, we have that

(a) $$J_0(x) = \left(\frac{1}{x}\frac{d}{dx}\right)^n [x^n J_n(x)],$$

and

(b) $$J_n(x) = (-1)^n \,.\, x^n \left(\frac{1}{x}\frac{d}{dx}\right)^n [J_0(x)].$$

Proof

(a) The recurrence relation (7.29) gives that

$$\frac{d}{dx}[x^n J_n(x)] = x^n J_{n-1}(x),$$

i.e.,

$$\left(\frac{1}{x}\frac{d}{dx}\right)[x^n J_n(x)] = x^{n-1} J_{n-1}(x). \tag{7.30}$$

Operating the function $x^n J_n(x)$ with the operator $\frac{1}{x}\frac{d}{dx}$, *n*-times, we deduce the result (a).

(b) The result (b) can be proved in a similar manner, if we express the recurrence relation (7.28), in the form :

$$\left(\frac{1}{x}\frac{d}{dx}\right)[x^{-n} J_n(x)] = (-1)\,x^{-(n+1)} J_{n+1}(x). \tag{7.31}$$

Using the operator $\frac{1}{x}\frac{d}{dx}$, k-times, we get

$$\left(\frac{1}{x}\frac{d}{dx}\right)^k [x^{-n} J_n(x)] = (-1)^k x^{-(n+k)} J_{n+k}(x). \tag{7.32}$$

Setting $n = 0$, in (7.32) we obtain

$$J_k(x) = (-1)^k x^k . \left(\frac{1}{x}\frac{d}{dx}\right)^k [J_0(x)] \tag{7.33}$$

and this proves the result (b), if k is replaced by n.

7.1.3 Various Representations of $J_n(x)$

In this section, we shall establish various important representations of the Bessel function of the first kind, $J_n(x)$. We start with the 'generating function' representation first and state the following theorem :

Theorem 7.1.8

For all real values of x and for values of t such that $0 < |t| < \infty$, the function $w(x, t) = \exp\left\{\frac{1}{2}x\left(t - \frac{1}{t}\right)\right\}$, generates the Bessel functions $J_n(x)$, of integral order n, i.e.,

$$\exp\left\{\frac{1}{2}x\left(t - \frac{1}{t}\right)\right\} = \sum_{n=-\infty}^{\infty} J_n(x)\, t^n \tag{7.34}$$

Proof

Using the standard expansion formulae for the exponential functions, we write

$$w(x, t) = \left(\sum_{r=0}^{\infty} \frac{(1/2\, xt)^r}{r!}\right)\left(\sum_{s=0}^{\infty} \frac{(-1/2\, x)^s\, t^{-s}}{s!}\right)$$

$$= \sum_{r=0}^{\infty}\sum_{s=0}^{\infty} \frac{(-1)^s}{\Gamma(r+1)\,\Gamma(s+1)} (1/2\, x)^{r+s} . t^{r-s}$$

i.e.

$$w(x, t) = \sum_{n=-0}^{\infty} f_n(x)\, t^n \quad \text{(say)} \quad (n = r - s) \tag{7.35}$$

where $f_n(x)$ is determined as follows :

Case (i) : $n \geq 0$

For a fixed value of 'r', t^n occurs in the above 'double' series, if $s = r - n$, and then the corresponding coefficient of t^n is given by

$$\frac{(-1)^{r-n}}{\Gamma(r+1)\,\Gamma(r-n+1)} (1/2\, x)^{2r-n}\ (r = 0, 1, 2, \ldots)$$

so that we obtain, for $n \geq 0$,

$$f_n(x) = \sum_{r=0}^{\infty} \frac{(-1)^{r-n}}{\Gamma(r+1)\,\Gamma(r-n+1)} (1/2\,x)^{2r-n}. \qquad (7.36)$$

Using the property of the gamma function that $\Gamma(m) \to 0$, for $m = 0, -1, -2, \ldots$, we obtain from (7.36) that, for $n \geq 0$

$$f_n(x) = \sum_{r=n}^{\infty} \frac{(-1)^{r-n}}{\Gamma(r+1)\,\Gamma(r-n+1)} (1/2\,x)^{2r-n}$$

$$= \sum_{p=0}^{\infty} \frac{(-1)^p}{\Gamma(n+p+1)\,\Gamma(p+1)} (1/2\,x)^{n+2p} \quad (\text{setting } r-n=p)$$

i.e.,

$$f_n(x) = J_n(x), \quad \text{for } n \geq 0, \qquad (7.37)$$

by using (7.11).

Case (ii) : $n < 0$

In this case, $f_n(x)$ is the same series as given by (7.36), with the understanding that $n < 0$, and we find, by using (7.11) again, that

$$f_n(x) = (-1)^{-n} J_{-n}(x) = J_n(x), \quad \text{for } n > 0, \qquad (7.38)$$

if the result (7.12) is also made use of. The theorem follows, if the two results (7.37) and (7.38) are utilized in the relation (7.35).

As a direct utility of above theorem 7.1.8, we prove the following important representation theorem for the Bessel functions $J_n(x)$:

Theorem 7.1.9

For 'integral' values of the order n, the Bessel functions of the first kind has the following representation :

$$J_n(x) = \frac{1}{\pi} \int_0^{\pi} \cos(n\,\phi - x \sin\phi)\, d\,\phi \quad (n = 0, \pm 1, \pm 2, \ldots)$$

Proof

Writing $t = e^{i\,\phi}$ $(0 \leq \phi \leq \pi)$, in the representation (7.34), we obtain

$$\exp(ix \sin\phi) = \sum_{m=-\infty}^{\infty} J_m(x)\, \exp(im\,\phi)$$

$$= J_0(x) + \sum_{m=1}^{\infty} \{J_m(x)\, e^{im\phi} + J_{-m}(x)\, e^{-im\phi}\}$$

$$= J_0(x) + \sum_{m=1}^{\infty} \{e^{im\phi} + (-1)^m e^{-im\phi}\}\, J_m(x)$$

(by using (7.12))

$$\exp(ix \sin \phi) = J_0(x) + 2 \sum_{k=1}^{\infty} J_{2k}(x) \cos(2k\phi)$$

$$+ 2i \sum_{k=0}^{\infty} J_{2k+1}(x) \sin((2k+1)\phi),$$

(separating the even and odd parts of the series).

Equating the real and imaginary parts from both sides of the above relation, we arrive at the following two identities :

$$\text{(i)} \quad \cos(x \sin \phi) = J_0(x) + 2 \sum_{k=1}^{\infty} J_{2k}(x) \cos(2k\phi)$$

and

$$\text{(ii)} \quad \sin(x \sin \phi) = 2 \sum_{k=1}^{8} J_{2k+1}(x) \sin((2k+1)\phi).$$

Multiplying both sides of the identity (i) by $\cos(n\phi)$ and integrating w.r.t. ϕ, between the limits $\phi = 0$ and $\phi = \pi$, (here $n = a$ positive integer), we get

$$\int_0^{\pi} \cos(x \sin \phi) \cos(n\phi)\, d\phi = J_0(x) \int_0^{\pi} \cos(n\phi)\, d\phi$$

$$+ 2 \sum_{k=1}^{\infty} \int_0^{\pi} \cos(n\phi) \cos(2k\phi)\, d\phi\, J_{2k}(x), \tag{7.39}$$

the term by term integration being a permissible step because of the uniform convergence of the series on the right of (i).

We next use the result that (the orthogonality relation) :

$$\int_0^{\pi} \cos(n\phi) \cos(p\phi)\, d\phi = \begin{cases} 0, & n \neq p \ (p = a \text{ positive integer}) \\ \pi/2, & n = p \end{cases} \tag{7.40}$$

and obtain, from (7.39), that

$$\int_0^{\pi} \cos(x \sin \phi) \cos(n\phi)\, d\phi = \begin{cases} \pi J_n(x), & \text{if } n \text{ is even} \\ 0, & \text{if } n \text{ is odd} \end{cases} \tag{7.41}$$

In a similar fashion, we obtain from the identity (ii), the result that

$$\int_0^{\pi} \sin(x \sin \phi) \sin(n\phi)\, d\phi = \begin{cases} \pi J_n(x), & \text{if } n \text{ is odd} \\ 0, & \text{if } n \text{ is even,} \end{cases} \tag{7.42}$$

if the following orthogonality relation is made use of:

$$\int_0^{\pi} \sin(n\phi) \sin(p\phi)\, d\phi = \begin{cases} 0, & \text{if } n \neq p \\ \pi/2, & \text{if } n = p. \end{cases} \tag{7.43}$$

Combining the two results (7.41) and (7.42), we, therefore, obtain the identity that

$$J_n(x) = \frac{1}{\pi}\int_0^{\pi} \cos(n\,\phi - x \sin\phi)\, d\,\phi, \tag{7.44}$$

for all positive integral values of n.

That the result (7.44) is also valid for $n=0$, can be proved by just integrating the relation (i), obtained above w.r.t. ϕ, between the limits $\phi = 0$ and $\phi = \pi$.

In order to prove that the relation (7.44) also holds good for $n = a$ negative integer, we proceed as follows :

We can write, for $k = a$ positive integer.

$$\frac{1}{\pi}\int_0^{\pi} \cos\{-k\,\phi - x\sin\phi\}\, d\,\phi$$

$$= \frac{1}{\pi}\int_0^{\pi} \cos\{-k(\pi - \theta) - x\sin\theta\}\, d\theta \quad (\text{setting } \phi = \pi - \theta).$$

$$= (-1)^k \frac{1}{\pi}\int_0^{\pi} \cos(k\,\theta - x\sin\theta)\, d\theta$$

$$= (-1)^k J_k(x), \qquad (\text{by using (7.44)}).$$

Thus, setting $k = -n$, and using the result (7.12), we obtain,

$$\frac{1}{\pi}\int_0^{\pi} \cos(n\,\phi - x\sin\phi)\, d\,\phi = (-1)^n \cdot (-1)^n J_n(x) = J_n(x),$$

$$(\text{if } n = a \text{ negative integer}) \tag{7.45}$$

This completes the proof of the theorem.

As a straightforward application of the integral representation theorem 7.1.9., we evaluate the following improper integral involving the Bessel function $J_0(x)$, in a closed form

$$I = \int_0^{\infty} e^{-ax} J_0(bx)\, dx \quad (a, b > 0).$$

Using the representation

$$J_0(x) = \frac{1}{\pi}\int_0^{\pi} \cos(x\sin\phi)\, d\phi,$$

we obtain after interchanging the orders of integration,

$$I = \frac{1}{\pi}\int_0^{\pi} d\phi \int_0^{\infty} e^{-ax} \cos(bx\sin\phi)\, dx.$$

Evaluating the inner integral by a standard technique, we get,

$$I = \frac{a}{\pi}\int_0^{\pi} \frac{d\phi}{a^2 + b^2 \sin^2 \phi} = \frac{a}{\pi}\int_0^{\pi} \frac{d\phi}{(a^2+b^2) - b^2\cos^2\phi}$$

$$= \frac{a}{\pi}\int_0^{\pi} \frac{\sec^2\phi \, d\phi}{a^2 + (a^2+b^2)\tan^2\phi}$$

$$= \frac{a}{\pi}\int_{-\pi/2}^{\pi/2} \frac{\operatorname{cosec}^2\phi \, d\phi}{a^2 + (a^2+b^2)\cot^2\phi} \quad \text{(chainging } \phi \text{ to } \pi/2 - \phi)$$

$$= \frac{2a}{\pi}\int_0^{\pi/2} \frac{\operatorname{cosec}^2\phi \, d\phi}{a^2 + (a^2+b^2)\cot^2\phi}$$

$$= \frac{2a}{\pi(a^2+b^2)}\sqrt{\frac{a^2+b^2}{a^2}}\,\frac{\pi}{2} = \frac{1}{\sqrt{a^2+b^2}}.$$

An integral representation of the Bessel function $J_n(x)$ that is valid for more general values of the 'order' n, is supplied by the following theorem :

Theorem 7.1.10

For values of n for which $Re\,(n) > -1/2$, we have that

$$J_n(x) = \frac{(x/2)^n}{\sqrt{\pi}\,\Gamma\left(n + \frac{1}{2}\right)} \int_{-1}^{1} (1 - t^2)^{n - 1/2} e^{ixt}\, dt \qquad (7.46)$$

Proof

We observe that the integral on the right of equation (7.46) can be expressed in the form :

$$I = \int_{-1}^{1} (1 - t^2)^{n-1/2} e^{ixt}\, dt$$

$$= \sum_{s=0}^{\infty} \frac{(ix)^s}{s\,!} \int_{-1}^{1} (1 - t^2)^{n-1/2} t^s\, dt,$$

by using the expansion for e^{ixt} and interchanging the orders of integration and summation.

Observing that t^s is an odd function of t, if s is odd, we obtain

$$I = 2\sum_{r=0}^{\infty} \frac{(+x)^{2r}(-1)^r}{(2r)\,!} \int_0^1 (1 - t^2)^{n-1/2} t^{2r}\, dt$$

$$= \sum_{r=0}^{\infty} \frac{(-1)^r x^{2r}}{(2r)\,!} \int_0^1 (1 - u)^{n-1/2} u^{r-1/2}\, du \qquad (\text{setting } t^2 = u)$$

$$I = \sum_{r=0}^{\infty} \frac{(-1)^r x^{2r}}{(2r)\,!} B\left(n + \frac{1}{2}, r + \frac{1}{2}\right) \left(Re\,(n) > -\frac{1}{2}\right) \qquad (7.47)$$

where $B(m, n)$ is the Beta function.

We now use of identities that

$$\text{(i)} \quad B(m, n) = \frac{\Gamma(m)\,\Gamma(n)}{\Gamma(m+n)},$$

and

$$\text{(ii)} \quad \Gamma\left(r+\frac{1}{2}\right) = \left(r-\frac{1}{2}\right)\Gamma\left(r-\frac{1}{2}\right) = \left(r-\frac{1}{2}\right)\left(r-\frac{3}{2}\right)\Gamma\left(r-\frac{3}{2}\right)$$

$$= \ldots = \left(r-\frac{1}{2}\right)\left(r-\frac{3}{2}\right)\ldots\frac{1}{2}\Gamma\left(\frac{1}{2}\right)$$

$$= \frac{1}{2^{2r}}\frac{(2r)\,!}{r\,!}\sqrt{\pi}. \qquad (\because\ \Gamma(1/2) = \sqrt{\pi}).$$

Then we find, from (7.47), that

$$I = \Gamma\left(n+\frac{1}{2}\right)\sqrt{\pi}\sum_{r=0}^{\infty}\frac{(-1)^r}{\Gamma(r+1)\,\Gamma(n+r+1)}(x/2)^{2r}$$

$$= \Gamma\left(n+\frac{1}{2}\right)\sqrt{\pi}\,(x/2)^{-n}\,J_n(x), \quad \text{for } Re(n) > -\frac{1}{2}$$

(by using (7.11))

and the theorem follows immediately. Making the substitution $t = \cos\theta$, in (7.46), we obtain

$$J_n(x) = \frac{(x/2)^n}{\Gamma(1/2)\,\Gamma(n+1/2)}\int_0^{\pi}\cos(x\cos\theta)\sin^{2n}\theta\,d\theta, \tag{7.48}$$

for $Re(n) > -\frac{1}{2}$.

and this formula (7.48) serves as an integral representation for the function $J_n(x)$, for $Re(n) > -\frac{1}{2}$.

For arbitrary values of n, we have the following representation theorem.

Theorem 7.1.11

For all values of n, we have that

$$J_n(x) = \frac{1}{\pi}\int_0^{\pi}\cos(x\sin\theta - n\theta)\,d\theta - \frac{\sin n\pi}{\pi}$$

$$\int_0^{\infty}\exp(-x\sinh\alpha - n\,\alpha\,d\,\alpha)\ (x \text{ real}). \tag{7.49}$$

(*Note :* Taking n to be an integer, theorem 7.1.9 follows).

Proof

We make use of the formula (see Levedev [8])

$$\frac{1}{\Gamma(z)} = \frac{1}{2\pi i}\int_c e^s s^{-z}\, ds, \tag{7.50}$$

where c is a positively oriented contour consisting of two straight lines, parallel to the negative real axis, one just above and the other just below the real axis, as well as a circular loop around the origin of the complex s-plane. Using the formula (7.50) in the definition (7.11) of $J_n(x)$, we obtain

$$J_n(x) = \sum_{k=0}^{\infty} \frac{(-1)^k (x/2)^{n+2k}}{\Gamma(k+1)} \frac{1}{2\pi i}\int_c e^s (s^{-(n+k+1)}\, ds$$

$$= (x/2)^n \frac{1}{2\pi i}\int_c e^s s^{-\nu-1}\, ds \left(\sum_{k=0}^{\infty} \frac{(-1)^k (x^2/4s)^k}{\Gamma(k+1)}\right)$$

(interchanging the orders of integration and summation, valid because of uniform convergence)

$$= (x/2)^n \frac{1}{2\pi i}\int_c e^{x - x^2/4s}\, s^{-n-1}\, ds.$$

Now, observing that x is real, we set $s = xt/2$, and obtain

$$J_n(x) = \frac{1}{2\pi i}\int_{c'} \exp\left\{\frac{1}{2}x\left(t - \frac{1}{t}\right)\right\} t^{-n-1}\, dt, \tag{7.51}$$

where c' is a contour in the complex t-plane, which is similar to the contour c. (Note that the formula (7.51) can be shown to be valid, even for complex values of x, for which $|\arg x| < \pi/2$, if we use the principle of analytic continuation.)

Writing $t = \rho e^{i\theta}$, and choosing the radius of the circular part of c' as 1, we have

$$J_n(x) = \frac{1}{\pi}\int_0^{\pi} \cos(x\sin\theta - n\theta)\, d\theta - \frac{\sin n\pi}{\pi}$$

$$-\int_1^{\infty} \exp\left\{-\frac{1}{2}x(\rho - \rho^{-1})\,.\,\rho^{-n-1}\right\} d\rho. \tag{7.52}$$

Setting $\rho = e^x$ in the second integral above, the theorem follows.

7.2 SOME IMPORTANT INTEGRALS INVOLVING $J_n(x)$

In this section we shall evaluate some important integrals involving the Bessel function $J_n(x)$, as these occur quite naturally in the studies of Boundary value problems of Mathematical Physics (see Sneddon) [12], [14]),

The integral $I_1 = \int_0^a x^n J_{n-1}(x)\,dx, (a > 0, Re\,(n) > 0)$.

Using the recurrence relation (iv) of Theorem 7.1.6, we observe that

$$I_1 = \int_0^a \frac{d}{dx}[x^n J_n(x)]\,dx = a^n J_n(a). \tag{7.53}$$

Setting $x = \xi y$ and $a = \xi\alpha$, we deduce that

$$\int_0^\alpha y^n J_{n-1}(\xi\, y)\,dy = \frac{\alpha^n}{\xi} J_n(\xi\,\alpha), \tag{7.54}$$

and, in particular, choosing $n = 1$, we get

$$\int_0^\alpha yJ_0(\xi\, y)\,dy = \frac{\alpha}{\xi} J_1(\xi\,\alpha),\ (\alpha > 0). \tag{7.55}$$

.. The integral $I_2 = \int_0^a x^{n-2} J_{n-1}(x)\,dx,\ (a > 0, Re\,(n) > 0)$.

We write

$$I_2 = \int_0^a x^2\,[x^n J_{n-1}(x)]\,dx,$$

and obtain, by using the recurrence relation (iv) of Theorem 7.1.16, again, that

$$I_2 = \int_0^a x^2 \frac{d}{dx}[x^n J_n(x)]\,dx$$

$$= a^{n+2} J_n(a) - 2\int_0^a x^{n+1} J_n(x)\,dx, \quad \text{(integrating by parts)}$$

or,

$$I_2 = a^{n+2} J_n(a) - 2\int_0^a \frac{d}{dx}[x^{n+1} J_{n+1}(x)]\,dx,$$

(using (iv) of Theorem 7.1.6 again)

i.e.,

$$I_2 = a^{n+2} J_n(a) - 2a^{n+1} J_{n+1}(a). \tag{7.56}$$

Setting $x = \xi y$, $\alpha = a/\xi$, in (7.56) we deduce that

$$\int_0^\alpha y^{n+2} J_{n-1}(\xi\, y)\,dy = \frac{2\,\alpha^{n+1}}{\xi^2}\left[J_{n-1}(\alpha\,\xi) + \left(\frac{\alpha\,\xi}{2} - \frac{2n}{\alpha\,\xi}\right)J_n(\alpha\,\xi)\right]. \tag{7.57}$$

In particular, choosing $n = 1$ in (7.57), we obtain

$$\int_0^{\alpha} x^3 J_0(\xi x)\, dx = \frac{2\alpha^2}{\xi^2}\left[J_0(\alpha\xi) + \left(\frac{\alpha\xi}{2} - \frac{2}{\alpha\xi}\right) J_1(\alpha\xi)\right] (\alpha > 0). \quad (7.58)$$

Finite integrals involving the Bessel function $J_n(x)$ and the powers of x of the type x^{m+n} (m = integer) can be evaluated in a similar manner.

3. The integral $I_3 = I(n, \lambda, \mu) = \int_0^a x J_n(\lambda x) J_n(\mu x)\, dx$, ($a > 0$, λ, μ = real, $Re(n) > -1/2$).

In order to evaluate the integral $I(n, \lambda, \mu)$, we proceed as follows.

Observe that $J_n(\lambda x)$ and $J_n(\mu x)$ satisfy the corresponding Bessel equation; i.e.

$$\left[x^2 \frac{d^2}{dx^2} + x\frac{d}{dx} + (\lambda^2 x^2 - n^2)\right] J_n(\lambda x) = 0 \quad \text{(i)}$$

and

$$\left[x^2 \frac{d^2}{dx^2} + x\frac{d}{dx} + (\mu^2 x^2 - n^2)\right] J_n(\mu x) = 0 \quad \text{(ii)}$$

(i) $\times J_n(\mu x)$ – (ii) $\times J_n(\lambda x)$ gives that

$$(\mu^2 - \lambda^2)\, x J_n(\lambda x) J_n(\mu x) = \frac{d}{dx}\left\{x\left[J_n(\mu x)\frac{dJ_n(\lambda x)}{dx} - J_n(\lambda x)\frac{dJ_n(\mu x)}{dx}\right]\right\}.$$

Integrating both sides w.r.t. x, between $x = 0$ and $x = a$, and using the condition that $Re(n) > -1/2$ we thus obtain that

$$I(n, \lambda, \mu) = \frac{a}{(\mu^2 - \lambda^2)}[\lambda J_n(\mu a) J'_n(\lambda a) - \mu J_n(\lambda a) J'_n(\mu a)] \quad (7.59)$$

if $\lambda \neq \mu$, dash denoting differentiation with respect to the argument. From the result (7.59), we immediately infer that if 'λa' and 'μa' are the roots of the transcendental equation $J_n(x) = 0$ (It can be shown that for values of $n = 0, 1, 2, \ldots$, the roots 'λa' and 'μa' are real (see Watson [15]) and a discussion regarding these roots is beyond the scope of this book), then we have that

$$\int_0^a x J_n(\lambda x) J_n(\mu x)\, dx = 0, \quad \text{for } \lambda \neq \mu. \quad (7.60)$$

In order to evaluate the integral $\int_0^a x\{J_n(\lambda x)\}^2 dx$, in the case when '$\lambda a$' is a root of $J_n(x) = 0$, we employ L' Hospital's rule in the result (7.59) and obtain

$$\int_0^a x\{J_n(\lambda x)\}^2 dx = \lim_{\lambda \to \mu} I(n, \lambda, \mu)$$

$$= \frac{a^2}{2}\{J'_n(\lambda a)\}^2 - \frac{a^2}{2} J_n(\lambda a) J''_n(\lambda a)$$

$$= \frac{a^2}{2}\{J'_n(\lambda a)\}^2 = \frac{a^2}{2}\{J_{n+1}(\lambda a)\}^2, \tag{7.61}$$

(by using the recurrence relation (i), of Theorem 7.1.6, since $J_n(\lambda a) = 0$).

The above two results (7.60) and (7.61) provide us with the following orthogonality relations for the Bessel functions of the first kind :

$$\int_0^a x\, J_n(\lambda_i x)\, J_n(\lambda_j x)\, dx = \frac{a^2}{2}\{J_{n+1}(\lambda_i a)\}^2\, \delta_{ij}, \tag{7.62}$$

where δ_{ij} is the Krönecker delta and $\{\lambda_i\}$ $(i = 1, 2, ...)$ is an ordered set of zeros of the Bessel function $J_n(\lambda a) = 0$, for a fixed value of the positive number 'a' and for $n = 0, 1, 2, ...$

The utility of the above orthogonality relation (7.62) is that a certain class of functions $f(x)$, defined in the interval $[0, a]$, can be expanded in the form of a convergent series, known as the Fourier-Bessel series, and this expansion formula is given by

$$f(x) = \sum_{m=1}^{\infty} c_m J_n\left(\lambda_{mn}\frac{x}{a}\right),\quad 0 < x < a,\ (n = 0, 1, 2, ...), \tag{7.63}$$

where

$$c_m = \frac{2}{a^2 J_{n+1}^2(\lambda_{mn})}\int_0^a xf(x)\, J_n\left(\lambda_{nm}\frac{x}{a}\right)dx,\ (m = 1, 2, \ldots) \tag{7.64}$$

λ_{nm} being the roots of the equation $J_n(\lambda) = 0$.

Problems of evaluating definite integrals involving the Bessel function $J_n(x)$ can be handled by exploiting the various representations of $J_n(x)$ discussed here and a number of such integrals are included in the list of problems provided at the end of the chapter.

7.3 OTHER BESSEL FUNCTIONS

Using the knowledge about the two Bessel functions $J_n(x)$ and $Y_n(x)$ introduced and discussed before, we define two more Bessel functions as follows :

Definition 7.1.4

The two 'special' functions $H_n^{(1)}(x)$ and $H_n^{(2)}(x)$ as given by

$$H_n^{(1)}(x) = J_n(x) + iY_n(x) \tag{7.65}$$

and

$$H_n^{(2)}(x) = J_n(x) - iY_n(x), \tag{7.66}$$

are called the Bessel functions of third and fourth kind respectively. These two functions are also known as the Hankel functions of the first and second kind respectively. It is easily verified that the Hankel functions are two independent solutions of the Bessel equation (7.1). Using the representations for the functions $J_n(x)$ and $Y_n(x)$, as given by (7.11) and (7.14a, b), we can easily show that the following results hold:

$$\left.\begin{aligned}
&\text{(i)}\quad H^{(1)}_{-n}(x) = e^{n\pi i}\, H^{(1)}_n(x),\\
&\text{(ii)}\quad H^{(2)}_{-n}(x) = e^{-n\pi i}\, H^{(2)}_n(x),\\
&\text{(iii)}\quad 2J_{-n}(x) = e^{n\pi i}\, H^{(1)}_n(x) + e^{-n\pi i}\, H^{(2)}_n(x),\\
&\text{(iv)}\quad 2iY_{-n}(x) = e^{n\pi i}\, H^{(1)}_n(x) - e^{-n\pi i}\, H^{(2)}_n(x).
\end{aligned}\right\} \tag{7.67}$$

A differential equation that is closely related with the Bessel equation (7.1) is the differential equation given by

$$x^2 \frac{d^2 y}{dx^2} + x \frac{dy}{dx} - (x^2 + n^2)\, y = 0, \tag{7.68}$$

and this equation is known as the 'Modified Bessel Equation'.

By changing the variable x in equation (7.11) to the new variable ix, with $i^2 = -1$, it is easily verified that equation (7.68) results and, therefore, by a similar change of the variable x in the definitions of the various Bessel functions introduced earlier, we define,

Definition 7.1.5

The two special functions, as given by

$$I_n(x) = e^{-1/2\, n\pi i}\, J_n(xe^{i\pi/2}) \tag{7.69}$$

and

$$K_n(x) = \frac{1}{2}\pi \frac{\{I_{-n}(x) - I_n(x)\}}{\sin n\pi} \qquad (n \neq \text{an integer}) \tag{7.70}$$

and

$$K_n(x) = \lim_{\nu \to n} K_\nu(x) \qquad (n = \text{an integer})$$

are called the 'Modified Bessel Functions' of the 'first' and 'second' kind respectively.

It is easily verified that the two functions $I_n(x)$ and $K_n(x)$ represent two linearly independent solutions of equation (7.68).

From the various definitions and properties of the Bessel functions introduced earlier, we can easily deduce that the following relations hold good:

(i) $I_{-n}(x) = I_n(x)$, (n = an integer)

(ii) $K_{-n}(x) = K_n(x)$, for all n,

(iii) $K_n(x) = \dfrac{\pi i}{2} \exp\left(\dfrac{n\pi i}{2}\right) H^{(1)}_n(xe^{\pi i/2})$,

(iv) $K_n(xe^{i\pi/2}) = -\dfrac{\pi i}{2} \exp\left(-\dfrac{n\pi i}{2}\right) H^{(2)}_n(x)$.

The following recurrence relations also can be proved easily:

(i) $I_{n-1}(x) - I_{n+1}(x) = \dfrac{2n}{x} I_n(x)$,

(ii) $I_{n-1}(x) + I_{n+1}(x) = 2I'_n(x)$,

(iii) $xI'_n(x) + nI_n(x) = xI_{n-1}(x)$,

(iv) $xI'_n(x) - nI_n(x) = xI_{n+1}(x)$,

(v) $K_{n-1}(x) - K_{n+1}(x) = -\dfrac{2n}{x} K_n(x)$,

(vi) $K_{n-1}(x) + K_{n+1}(x) = -2K'_n(x)$

(vii) $xK_n(x) + nK_n(x) = -xK_{n-1}(x)$

and

(viii) $xK'_n(x) - nK_n(x) = -xK_{n+1}(x)$.

7.4 THE WRONSKIAN

The calculation of the Wronskians of two independent solutions of the Bessel equation (7.1) and those of the modified Bessel equation (7.68) are important problems as these have wide applications in boundary value problems of Mathematical Physics. We give below a method of determining the value of the Wronskian in connection with the Bessel equation (7.1), while a similar method can be adopted for the modified equation (7.68).

If $y_1(x)$ and $y_2(x)$ are regarded as any two linearly independent solutions of the Bessel equation (7.1) we have that

$$x^2 y_1'' + x y_1' + (x^2 - n^2) y_1 = 0$$

and

$$x^2 y_2'' + x y_2' + (x^2 - n^2) y_2 = 0.$$

Multiplying the first equation by y_2 and the second by y_1 and taking difference, we obtain

$$\frac{d}{dx}[x(y_1' y_2 - y_1 y_2')] = 0,$$

and an integration gives that

$$y_1 y_2' - y_1' y_2 \equiv W(y_1, y_2) = \frac{c}{x}, \tag{7.71}$$

where $W(y_1, y_2)$ represents the Wronskian of the two solutions y_1 and y_2, and c is an arbitrary constant.

By using the result (7.71), in conjunction with the series representations of the various Bessel functions, we find that

(i) $W[J_n(x), J_{-n}(x)] = -2\sin(n\pi)/(\pi x)$,

(ii) $W[J_n(x), Y_n(x)] = 2/(\pi x)$,

and

(iii) $W[H_n^{(1)}(x), H_n^{(2)}(x)] = -4i/(\pi x)$.

Similar results can be deduced for various combinations of the other Bessel functions, and, for the modified Bessel functions $I_n(x)$ and $K_n(x)$. We find that the following relations hold:

$$(*)\quad W[I_n(x),\ I_{-n}(x)] = -2\sin(n\pi)/(\pi x),$$

and

$$(**)\quad W[K_n(x),\ I_n(x)] = 1/x$$

Further properties of the Bessel functions are not included in the present book, the purpose of which lies mainly in introducing the subject to the beginners. However, the problems listed afterwards include some such properties which have not been discussed in the main text.

PROBLEMS

1. Show that $J'_0(x)$ satisfies the Bessel equation of order one.

2. Show that $J_0(0) = 1$ and $J_0(x)$ is continuous in some interval $0 < x < a$, for some $a > 0$, in which $J_0(x) \neq 0$.

Prove that if $0 < x_0 < a$, then the second solution of the Bessel equation of order zero can be taken in the form

$$y_2(x) = J_0(x)\int_{x_0}^{x}\left[\frac{1}{tJ_0^2(t)}\right]dt,$$

$y_1(x) = J_0(x)$ being the first solution.

3. Prove the following relations:

(i) $8J_n'''(x) = J_{n-3}(x) - 3J_{n-1}(x) + 3J_{n+1}(x) - J_{n+2}(x),$

(ii) $4J_0'''(x) + 3J_0'(x) + J_3(x) = 0.$

4. Using the expression for $P_n(\cos\theta)$, or otherwise, prove that

$$\sum_{n=0}^{\infty}\frac{r^n}{n!}P_n(\cos\theta) = e^{r\cos\theta}J_0(r\sin\theta).$$

5. Prove the following results, for $a > 0, p > 0, Re(\nu) > -\frac{1}{2}$.

(i) $\displaystyle\int_0^\infty J_\nu(ax)\,x^\nu e^{-px}\,dx = \frac{2^\nu\,\Gamma(\nu+1/2)}{\Gamma(1/2)}\,\frac{a^\nu}{(a^2+p^2)^{\nu+1/2}}.$

(ii) $\displaystyle\int_0^\infty J_\nu(ax)\,x^{\nu+1}e^{-px}\,dx = \frac{2^{\nu+1}\,\Gamma(\nu+3/2)}{\Gamma(1/2)}\,\frac{pa^\nu}{(a^2+p^2)^{\nu+3/2}}$

6. Using the series expansion for $J_n(x)$, prove the following result (known as Sonine's first finite integral):

$$J_{\mu+\nu+1}(x) = \frac{x^{\nu+1}}{\Gamma(\nu+1)\,2^\nu}\int_0^{\pi/2} J_\mu(x\sin\phi)\sin^{\mu+1}\phi\,\cos^{2\nu+1}\phi\,d\phi,$$

valid for $Re\,\mu > -1$, $Re\,\nu > -1$.

7. Prove that if a and b are real, and $Re\,\nu > -1$,

$$\int_0^\infty J_\nu(at)\, t^{\nu+1} \exp(-b^2 t^2)\, dt = \frac{a^\nu}{(2b^2)^{\nu+1}} \exp(-a^2/4b^2).$$

8. Prove that

$$\int_0^1 u^3 J_0(u)\, du = J_1(1) - 2 J_2(1).$$

9. Prove that

$$\sum_{n=-\infty}^{\infty} J_n(kx)\, t^n = \exp\left\{-\frac{x}{2t}\left(k - \frac{1}{k}\right)\right\} \sum_{n=-\infty}^{\infty} k^n t^n J_n(x),$$

and deduce that

(i) $J_n(re^{i\theta}) = \sum_{m=0}^{\infty} J_{n+m}(r)\, e^{i(n+m)\theta} \dfrac{(-ir\sin\theta)^m}{m\,!}$,

(ii) $I_n(x) = \sum_{m=0}^{\infty} \dfrac{x^m}{m\,!} J_{n+m}(x)$.

10. Using the expansion formula of problem 9, prove that

$$\exp\left\{-\frac{i(a\sin\alpha + b\sin\beta)}{t}\right\} \sum_{m=-\infty}^{\infty} \sum_{n=-\infty}^{\infty} e^{in\alpha - im\beta} J_n(a)\,.\,J_m(b)\, t^{n+m}$$

$$= \sum_{r=-\infty}^{\infty} t^r J_r(ae^{i\alpha} + be^{i\beta}).$$

Setting $R = a\cos\alpha + b\cos\beta$, $0 = a\sin\alpha + b\sin\beta$, and $\beta = \alpha + \theta - \pi$, prove Neumann's addition theorem for the Bessel functions $J_n(x)$, in the form :

$$J_n(R) = \left(\frac{a - be^{-i\theta}}{a - be^{i\theta}}\right)^{n/2} \sum_{m=-\infty}^{\infty} J_{n+m}(a)\, J_m(b)\, e^{im\theta},$$

where $R^2 = a^2 + b^2 - 2ab\cos\theta$.

11. If $\lambda_1, \lambda_2, \ldots$ are the positive zeros of $J_0(x)$, show that,

$$-\frac{1}{2}\log x = \sum_{i=1}^{\infty} \frac{J_0(\lambda_i x)}{\lambda_i^2 J_1^2(\lambda_i)}, \text{ for } x > 0.$$

Multiplying both sides of this equation by x and integrating term by term, deduce that

$$\frac{1}{8}x(1-2\log x) = \sum_{i=1}^{\infty} \frac{J_1(\lambda_i x)}{\lambda_i^3 J_1^2(\lambda_i)}.$$

12. Prove that, if $Re(\nu) > -\frac{1}{2}$,

$$I_\nu(x) = \frac{(x/2)^\nu}{\Gamma(\nu+1/2)\,\Gamma(1/2)} \int_0^\pi \exp(\pm\, x\cos\phi)\sin^{2\nu}\phi\, d\phi$$

and that

$$K_\nu(x) = \int_0^\infty \exp(-x\cosh t)\cosh(\nu t)\, dt.$$

13. Show that the general solution of the equation

$$4\frac{d^2y}{dx^2} + 9xy = 0$$

can be taken in the form

$$y = Ax^{1/2} J_{1/3}(x^{3/2}) + Bx^{1/2} J_{-1/3}(x^{3/2}),$$

where A and B are two arbitrary constants.

14. Prove that

(i) $J_{1/2}(x) = (2/\pi x)^{1/2} \sin x,$

(ii) $J_{-1/2}(x) = (2/\pi x)^{1/2} \cos x,$

(iii) $J_{3/2}(x) = (2/\pi x)^{1/2} \left\{\dfrac{\sin x}{x} - \cos x\right\},$

(iv) $J_{-3/2}(x) = (2/\pi x)^{1/2} \left\{-\sin x - \dfrac{\cos x}{x}\right\}$

15. Show that

$$1 = \frac{2}{a}\sum_{n=1}^{\infty} J_0(\alpha_n r)/[\alpha_n J_1(\alpha_n a)], \quad a > 0,$$

where $\alpha_1, \alpha_2, \ldots$, are the positive roots of $J_0(\alpha a) = 0$.

CHAPTER - EIGHT

The Existence and Uniqueness Theory for Ordinary Differential Equations – A brief discussion

In this chapter we shall describe very briefly the theory of existence and uniqueness of solutions of Ordinary Differential Equations. Though, it is rather customary in any study of the theory of differential equations, to start with such existence and uniqueness theory, we have deliberately kept such a study pending so far in this book with the main purpose that we wanted to describe various methods and properties of the solutions of differential equations which are linear in nature first, in order to motivate ourselves to take up the study described in the present chapter. It is believed that the reader is now sufficiently well equipped with various methods of solution of linear differential equations and is rightly anxious to know about such an existence and uniqueness theory for more general differential equations.

8.1 THE FIRST ORDER EQUATION

We start with the following general problem associated with the first order ordinary differential equation:

Problem. Solve the O.D.E.

$$\frac{dy}{dx} = f(x, y) \tag{8.1}$$

with the 'condition' that $y = y_0$ when $x = x_0$ (such a 'condition' associated with first order differential equation is called 'the initial condition' of the given problem).

The question that one naturally wants to ask is 'For what class of functions $f(x, y)$, a solution of equation (8.1), satisfying the 'initial condition' $y = y_0$ when $x = x_0$, is (i) guaranteed first, and (ii) unique next?'.

Before answering this basic question for a sufficiently general class of function $f(x, y)$, we take up certain special examples to understand the above question.

Example 8.1. Solve

$$\frac{dy}{dx} = \begin{cases} y\,(1-2x), & \text{when } x > 0 \\ y\,(2x-1), & \text{when } x < 0, \end{cases}$$

with the 'initial condition', $y = 1$, when $x = 1$.

Solution

Rewriting the given equation as:

$$\frac{dy}{y} = \begin{cases} 1-2x, & \text{when } x > 0 \\ 2x-1, & \text{when } x < 0 \end{cases} \tag{8.2}$$

a straightforward integration gives that

$$y = \begin{cases} ce^{x-x^2}, & x > 0 \\ De^{x^2-x}, & x < 0 \end{cases} \tag{8.3}$$

where c and D are arbitrary constants.

The given initial condition $y = 1$, when $x = 1$ will be satisfied if we choose $c = 1$, and then choosing $D = 1$ also as required by the continuity of y (at $x = 0$), we observe that the function

$$y = \begin{cases} e^{x-x^2}, & x \geq 0 \\ e^{x^2} - x, & x \leq 0 \end{cases} \tag{8.4}$$

represents the unique solution of our problem. An inference that can be drawn out of the above solution (8.4) of our problem is that we have been able to discover the continuous unique solution y, as given by equation (8.4), of the problem associated with the O.D.E.

$$\frac{dy}{dx} = f(x, y) \equiv \begin{cases} y\,(1-2x); & x > 0 \\ y\,(2x-1), & x < 0 \end{cases}$$

with the initial condition: $y = 1$, when $x = 1$, in which the function $f(x, y)$ is a discontinuous function of the two variable (x, y), the discontinuities lying along the line $x = 0$.

We thus observe that there exists a class of initial value problems associated with O.D.E. $\frac{dy}{dx} = f(x, y)$, for which there exist: a continuous solution $y(x)$ even if $f(x, y)$ is not continuous and that the solution is unique.

Example 8.2. Solve

$$\frac{dy}{dx} = \sqrt{|\,y\,|}$$

with the 'initial' condition: $y = 0$, when $x = 0$.

Solution

We find that the given O.D.E. can be rewritten as

$$\frac{dy}{dx} = \begin{cases} \sqrt{y}\,, & \text{for } y > 0 \\ \sqrt{-y}\,, & \text{for } y < 0 \end{cases}$$

and, then, a straightforward integration gives

$$y^{1/2} = \frac{x}{2} + c_1, \quad \text{for } y > 0, \ (\text{all } x)$$

and

$$-(-y)^{1/2} = \frac{x}{2} + c_2, \quad \text{for } y < 0 \ (\text{all } x).$$

The initial condition "$y = 0$, when $x = 0$", will be satisfied if we choose $c_1 = c_2 = 0$, and then, a solution of the given ODE is obtained as:

$$y = \begin{cases} -\dfrac{1}{4}x^2, & y < 0 \\ +\dfrac{1}{4}x^2, & y > 0 \end{cases}$$

It is also easily verified that $y = 0$ represents another possible solution of the given initial value problem, showing thereby that there does not exist a unique solution of the problem.

An observation on the above problem is that the function $f(x, y) = \sqrt{|y|}$, in this problem, represents a continuous function, and this is in contrast with the problem considered in Example 8.1, in which a discontinuous $f(x, y)$ produced a unique solution of the initial value problem.

Example 8.3. Solve the O.D.E.

$$\frac{dy}{dx} = f(x, y) = \begin{cases} \dfrac{4x^3 y}{x^4 + y^2}, & \text{when } (x, y) \neq (0, 0) \\ 0\,, & \text{when } (x, y) = (0, 0) \end{cases}$$

with the initial condition: $y = 0$, when $x = 0$.

Solution

We first observe that the function $f(x, y)$ for this problem is a continuous function of the two variables (x, y) and then, that we can express the given O.D.E. in the following 'exact' form :

$$-d(x^4/y) + dy = 0$$

giving, on integration.

$$-x^4/y + y = 2c^2 \text{ (say)}$$

where $2c^2$ is an arbitrary constant of integration.

We thus obtain that the function

$$y = c^2 - (c^4 + x^4)^{1/2}$$

where c is an arbitrary real constant, represents a solution of the given initial value problem, satisfying the initial condition $y = 0$, when $x = 0$.

We observe then, that the present problem possesses infinitely many solutions, even though the right-hand function $f(x, y)$ is a continuous function of the two variables x and y.

The above three examples are sufficient enough to motivate ourselves to the study of the class of functions $f(x, y)$ and the nature of the initial conditions, for which the unique solution of a given initial value problem is ensured.

We start with the following definition:

Definition 8.1

A function of two variables $f(x, y)$ is said to satisfy the Lipschitz condition with respect to the variable y, if f satisfies the inequality

$$|f(x, Y) - f(x, y)| < K\,|Y - y|, \tag{8.5}$$

where K is a finite positive constant.

In the examples of the functions $f(x, y)$ considered above, we find that the following holds good :

(i) $f(x, y)$ in Example 8.1 satisfies the Lipschitz condition w.r.t. y

(ii) $f(x, y)$ in Example 8.2, does not satisfy the Lipschitz condition w.r.t. y in any neighbourhood of the origin, since

$$\left|\frac{\sqrt{|Y|} - \sqrt{|y|}}{|Y - y|}\right| > \text{any constant, (finite)}$$

if Y and y are near zero.

(iii) $f(x, y)$ in Example 3, also does not satisfy the Lipschitz condition w.r.t. y in any neighbourhood of the origin, as can be seen from the following observation :

Taking $y = px^2$, $Y = qx^2$, where p, q are constants, we have that

$$|f(x, Y) - f(x, y)| \equiv \left|\frac{4x^3\,(x^4 - Yy)}{(x^4 + y^2)\,(x^4 + Y^2)}\,(Y - y)\right|$$

$$= 4 \cdot \left|\frac{1 - pq}{(1 + p^2)\,(1 + q^2)}\right| \frac{|Y - y|}{|x|},$$

so that if p and q are chosen to satisfy the requirement that $pq \neq 1$, we obtain

$$\left|\frac{f(x, Y) - f(x, y)}{Y - y}\right| > \text{any finite constant,}$$

if x is near zero. Hence the Lipschitz condition w.r.t. y is violated near $x = 0$.

We are now in a position to state the precise existence and uniqueness theorem for first order Ordinary Differential Equations and this is as given below:

Theorem 8.1

Let M be the upper bound of the continuous function $|f(x, y)|$ in the domain D: $\{|x - x_0| \le a, |y - y_0| \le b\}$ and let $f(x, y)$ satisfy the Lipschitz condition w.r.t. y, in the domain D: $\{|x - x_0| \le h, |y - y_0| < b\}$, where $h = \text{Min}(a, b/M)$.

Then the initial value problem

$$\frac{dy}{dx} = f(x, y)$$

with the initial condition $y = y_0$, when $x = x_0$, possesses a unique solution for all values of x belonging to the interval I: $\{|x - x_0| \le h\}$.

We shall not take up the proof of this theorem in the present book and refer to Ince's book [6] for that purpose. Instead, we shall analyse the theorem in the light of the three examples that we have discussed earlier.

We observe the following :

(i) For the problem in Example 8.1, we had obtained a unique solution, mainly because $f(x, y)$ there satisfies the Lipschitz condition w.r.t. y.

(ii) For the problems in Examples 8.2 and 8.3, we had obtained two solutions and infinitely many solutions, respectively, mainly because the two $f(x, y)$ there do not satisfy the Lipschitz condition, and

(iii) Discontinuity in $f(x, y)$ did not prevent the existence of a solution of the given initial value problem, as observed in Example 8.1, and, at the same time, continuity of $f(x, y)$ did not guaranteed the existence of a unique solution, as observed in Examples 8.2 and 8.3.

8.2 SYSTEM OF FIRST ORDER EQUATIONS

A generalization of the above existence and uniqueness theorem applicable to a system of m first order equations and the initial value problem associated with such a system is provided by the following general theorem which is also stated here without proof.

Theorem 8.2

Let the continuous functions $|f_i(x, y_1, y_2, \ldots, y_m)$, $(i = 1, 2, \ldots, m)$ satisfy the 'generalized Lipschitz condition as given by

$$|f_i(x, Y_1, Y_2, \ldots, Y_m) - f_i(x, y_1, y_2, \ldots, y_m)|$$
$$< K_1 |Y_1 - y_1| + K_2 |Y_2 - y_2| + \ldots + K_m |Y_m - y_m|,$$

(K_i's are finite positive constants), for the points $(x, y_1, \ldots, y_m)$ and $(x, Y_1, \ldots Y_m)$ belonging to the domain D' : $\{ |x - x_0| < h (= \text{Min}(a, b_1/M, b_2/M, \ldots, b_m/M))$, $|y_i - y_i^0| < b_i\}$ (M = the greatest of the upperbounds of $f_1, f_2, \ldots, f_m$ in D : $\{ |x - x_0| < a, |y_i - y_i^0| < b_i\}$).

Then the system of Ordinary Differential Equations

$$\frac{dy_i}{dx} = f_i(x, y_1, y_2, \ldots, y_m), \quad (i = 1, 2, \ldots m) \tag{8.6}$$

satisfying the initial conditions

$$y_i = y_i^0, \text{ when } x = x_0,$$

possesses a unique solution for all values of x belonging to the interval $I : \{ |x - x_0| \le h\}$.

In order to apply the above theorem 8.2 to a linear ordinary differential equation of order 'n', i.e., the equation.

$$p_0(x)\, y^{(n)} + p_1(x)\, y^{(n-1)} + \ldots + P_n(x)\, y = r(x), \tag{8.7}$$

we notice that this can be identified to be equivalent to the following system of n first order ordinary differential equations :

$$\frac{dy}{dx} = y_1,$$

$$\frac{dy_1}{dx} = y_2,$$

$$\frac{dy_2}{dx} = y_3,$$

....

$$\frac{dy_{n-2}}{dx} = y_{n-1},$$

$$\frac{dy_{n-1}}{dx} = \frac{r(x)}{p_0(x)} - \frac{p_n(x)}{p_0(x)}\, y - \frac{p_{n-1}(x)}{p_0(x)}\, y_1 \ldots - \frac{p_1(x)}{p_0(x)}\, y_{n-1}. \tag{8.8}$$

We shall not go into any further detail with the existence and uniqueness theory in the present book and, instead, we close this chapter with a list of problems to help the understanding of the topic discussed.

PROBLEMS

1. Show that the equation

$$\frac{dy}{dx} = y/x$$

with the initial condition $y = y_0$ when $x = x_0$, possesses a unique solution except in the neighbourhood of the line $x = 0$. Also show that the integral curve is given by

(i) $y = \dfrac{y_0}{x_0}\, x$, when $x_0 \neq 0$

and

(ii) $x = 0$, when $x_0 = 0$ and $y_0 \neq 0$.

[*Definition 1 :* A solution of a first order O.D.E. is also called an 'integral curve' of the O.D.E.].

[*Definition 2 :* A point (x_0, y_0) is said to be a 'Singular Point' of the first order ODE, $dy/dx = f(x, y)$, with the initial condition $y = y_0$, when $x = x_0$, if it does not possess a 'unique' solution, i.e., either (i) no solution exists, or (ii) more than one solution exists].

2. Show that the point $x_0 = 0 = y_0$ i.e., the 'origin' is the only 'singular point' of the equation in problem 1.

3. What are the singular points for the following ODES' ?

(i) $\dfrac{dy}{dx} = m\dfrac{y}{x}$ with $y = y_0$, when $x = x_0$ and m real,

(ii) $\dfrac{dy}{dx} = \dfrac{x+y}{x}$, $y = y_0$, when $x = x_0$.

(iii) $\dfrac{dy}{dx} = \dfrac{x+y}{x-y}$, $y = y_0$, when $x = x_0$.

4. Integrate the equation

$$\frac{dy}{dx} = \frac{a_1 x + b_1 y}{c_1 x + d_1 y}, \quad (a_1\, b_1\, c_1,\, d_1 \text{ constants})$$

by using the transformation $y = xv\,(x)$, and determine the behaviour of the integral curves in the neighbourhood of the origin.

(*Hints :* Get $v + x\dfrac{dv}{dx} = \dfrac{a_1 + b_1 v}{c_1 + d_1 v}$

i.e $\quad x\dfrac{dv}{dx} = \dfrac{a_1 + b_1 v}{c_1 + d_1 v} - v = -\dfrac{d_1 v^2 + (c_1 - b_1) v - a_1}{c_1 + d_1 v}$

i.e. $\quad \dfrac{(c_1 + d_1 v)}{d_1 v^2 + (c_1\, b_1) v - a_1}\, dv + \dfrac{dx}{x} = 0$

which can be integrated by standard methods. Observe that the following three important cases arise:

I. $(b_1 - c_1)^2 + 4a_1\, d_1 > 0$,

II. $(b_1 - c_1)^2 + 4a_1\, d_1 < 0$,

III. $(b_1 - c_1)^2 + 4a_1\, d_1 = 0$.)

[*Definition 3.* For the above equation $\dfrac{dy}{dx} = \dfrac{a_1 x + b_1 y}{c_1 x + d_1 y}$ and for the above three cases I, II, and III, we call the origin, which is a 'singular point' associated with this equation,

(i) A NODE, in case I, if $a_1\, d_1 - b_1\, c_1 < 0$

(ii) A SADDLE POINT, in case I, if $a_1\, d_1 - b_1\, c_1 > 0$

(iii) A FOCAL POINT, in Case II
and
(iv) A NODE, in case III].

5. Investigate the existence and uniqueness of solutions of the following equations near the origin:

(i) $y' = y^2$,

(ii) $x^2 y' = y$

(iii) $y' = -\dfrac{x + 2x^3}{y + 2y^3}$,

(iv) $xy' + y^2 = 0$ $[y' \equiv dy/dx]$.

6. Express the following systems of differential equations into equivalent first-order systems:

(i) $y_1^{(2)}(x) = a_1 y_1(x) + b_1 y_1^{(1)}(x) + c_1 y_2(x) + d_1 y_2^{(1)}(x)$,

$y_2^{(2)}(x) = a_2 y_1(x) + b_2 y_1^{(1)}(x) + c_2 y_2(x) + d_2 y_2^{(1)}(x)$,

($a_1, b_1, c_1, d_1, a_2, b_2, c_2, d_2$, are constants).

(ii) $y_1^{(3)}(x) + y_1(x)\, y_1^{(2)}(x) + a(x)\,[y_2(x) - \{y_1^{(1)}(x)^2\}] = 0$

$y_2^{(2)}(x) + b(x)\, y_1(x)\, y_2^{(1)}(x) = 0.$

7. If $\phi(x)$ is a solution of the equation

$$y^{(2)}(x) + a_1(x)\, y^{(1)}(x) + a_2(x)\, y(x) = 0,$$

and if $\phi(x) \neq 0$ for x belonging to an interval I, show that the function $\psi(x) = \phi'(x)/\phi(x)$ satisfies in I the Riccati equation

$$y^{(1)}(x) = -y^2(x) - a_1(x)\, y(x) - a_2(x)$$

and hence study the original equation for existence and uniqueness of solution, under appropriate initial conditions.

CHAPTER - NINE

Eigenvalue Problems – A Brief discussion

In this chapter we shall briefly discuss some of the elementary concepts in the theory of ordinary differential equations, involving a class of problems known as the 'Eigen-value Problems'. For a detailed study of this class of problems, the reader is referred to the monograph of Eastham [4].

9.1 THE EIGEN-VALUE PROBLEM

Let

$$Ly(x) = \lambda y(x) \tag{9.1}$$

be a given differential equation, where λ is a constant and for simplicity we assume that L is a second order linear differential operator as given by

$$L = a_0(x)\frac{d^2}{dx^2} + a_1(x)\frac{d}{dx} + a_2(x), \tag{9.2}$$

in which a_0, a_1, a_2 are continuous functions of x in a given closed interval $[a, b]$ with $a_0(x) \neq 0$ there.

Let us also assume that the following two conditions (known as the 'boundary conditions') are to be met with by the solution of equation (9.1).

$$\text{and} \quad \left.\begin{aligned} a_{11}y(a) + a_{12}y'(a) + b_{11}y(b) + b_{12}y'(b) = 0 \\ a_{21}y(a) + a_{22}y'(a) + b_{21}y(b) + b_{22}y'(b) = 0 \end{aligned}\right\} \tag{9.3}$$

(a_{ij}, b_{ij} are constants)

(*Note :* It is assumed that the two conditions in equation (9.3) are 'linearly independent', i.e., one is not obtainable from the other by multiplication by a constant. The mathematical condition for this independence is that the matrix

$$\begin{bmatrix} a_{11} & a_{12} & b_{11} & b_{12} \\ a_{21} & a_{22} & b_{21} & b_{22} \end{bmatrix}$$

must be of rank 2).

Definition 9.1.1

If there exist (i) a number λ_0 and (ii) a function $\psi(x)$ which is not *identically equal to zero* in $[a, b]$ (i.e. $\psi(x)$ is a *nontrivial solution*) such that equation (9.1) and the boundary condition (9.3) are satisfied when $\lambda = \lambda_0$ and $y = \psi(x)$, then λ_0 is said to be an 'Eigenvalue' and $\psi(x)$, the corresponding 'Eigenfunction' of the 'Eigenvalue-problem' as given by equations (9.1) and the condition (9.3).

The following theorem can be easily established by just employing the above definition 9.1.1.

Theorem 9.1.1

If ψ_1 and ψ_2 are 'eigenfunctions' corresponding to the eigenvalue λ_0, then so is the function $c_1\psi_1 + c_2\psi_2$, where c_1 and c_2 are constants, provided that $c_1\psi_1 + c_2\psi_2$ is not identically equal to zero in $[a, b]$.

Remark

There cannot be more than two linearly independent eigenfunctions, corresponding to an eigenvalue λ_0 of the above eigenvalue problem, because the second order equation (9.1) cannot have more than two linearly independent solutions.

To illustrate the above concepts, we consider the equation

$$-y''(x) = \lambda y(x), \quad (0 \le x \le \pi) \tag{9.4}$$

with the boundary conditions

$$y(o) = 0 = y(\pi). \tag{9.5}$$

Ii $\lambda \neq 0$, every solution of equation (9.4) is of the form

$$y(x) = A\cos\sqrt{\lambda}x + B\sin\sqrt{\lambda}x, \tag{9.6}$$

where A and B are constants.

The condition $y(o) = 0$ gives that $A = 0$ and the condition $y(\pi) = 0$ gives that

$$B\sin(\sqrt{\lambda}\pi) = 0, \tag{9.7}$$

which will be satisfied for non-zero values of B (this is important, because if $B = 0$ also, then (9.6) reduces to the trivial solution and we are not interested (see definition 9.1.1) in such solutions at all), if

$$\sin(\sqrt{\lambda}\pi) = 0, \tag{9.8}$$

The roots of equation (9.8) determine all the possible eigenvalues of the given problem, and we find that these eigenvalues are given by the relation

$$\lambda = n^2 \ (n = 1, 2, 3, \ldots) \tag{9.9}$$

Next, if $\lambda = 0$, we go back to equation (9.4) and observe that every solution, in this case, is of the form

$$y = A_0 x + B_0, \tag{9.10}$$

where A_0, B_0 are arbitrary constants, and that the boundary conditions (9.5) ensure that $A_0 = 0 = B_0$, giving only the trivial solution $y = 0$, of the above boundary value problem.

Thus, according to our definition 9.1.1. we conclude that $\lambda = 0$ is not an 'eigenvalue' and that the only possible eigenvalues are given by the relations (9.9).

We can also determine the 'eigenfunctions' $\psi_n(x)$ of the above eigenvalue problem, for a given 'eigenvalue' $\lambda_n (= n^2)$, by using the solution

$$y = \psi_n(x) = B_n \sin(nx), \tag{9.11}$$

and by suitably choosing the constants B_n for each n.

We state below an important theorem without proof (see Eastham [4]):

Theorem 9.1.2

These are the following two alternatives for the eigenvalue problem (9.1) and (9.3):

either (i) every complex number is an eigenvalue.

or, (ii) the eigenvalues form a countable set with no finite limit point.

9.2 THE STURM-LIOUVILLE EQUATIONS

A special class of eigenvalue problems, known as 'Self-adjoint' eigenvalue problems, has wide applications in problems of Mathematical Physics.

We define,

Definition 9.2.1

The eigenvalue problem described by equation (9.1) and the conditions (9.3), with the operator L as given by the relation (9.2), is said to be 'self-adjoint', if the relation

$$\int_a^b \overline{g(x)}\, Lf(x)\, dx = \int_a^b f(x)\, \overline{Lg(x)}\, dx \tag{9.12}$$

holds good, for all functions $f(x)$ and $g(x)$ which are twice continuously differentiable in the interval $[a, b]$, and which satisfy the boundary conditions (9.3). (A bar above a function indicates the complex conjugate of the function, in usual notations,)

It is easily proved, by integration by parts, that the relation (9.12) holds for the operator L in (9.2) if

(i) $\overline{Lg(x)} = \dfrac{d^2}{dx^2}(a_0 \overline{g}) - \dfrac{d}{dx}(a_1 \overline{g}) + a_2 \overline{g}$,

and (ii) $[f, g](b) = [f, g](a)$

where

$$[f, g](x) = f'(x)\, a_0(x)\, g(\overline{x}) - f(x)\, \{a_0 \overline{g}\}' + a_1 f \overline{g}.$$

The most important cases of the operator L, for which the above two results (i) and (ii) hold are those for which

$$L = -\frac{d}{dx}\left[p(x)\frac{d}{dx}\right] + q(x), \tag{9.13}$$

where p and q are real-valued functions, and the boundary conditions are as follows:

$$\left.\begin{array}{lll} \text{either} & \text{(a)} & a_1 y(a) + a_2 y'(a) = 0 \\ \text{and} & \text{(b)} & b_1 y(b) + b_2 y'(b) = 0 \end{array}\right\}, \tag{9.14}$$

or

$$\left.\begin{array}{lll} & (a)' & y(a) - y(b) = 0 \\ \text{and} & (b)' & p(a)\, y'(a) - p(b)\, y'(b) = 0 \end{array}\right\} \tag{9.15}$$

We define the following:

Definition 9.2.2

The boundary conditions (9.14) are called 'Separated' boundary conditions and the boundary conditions (9.15) are called 'Periodic' boundary conditions, for the operator (9.13).

Definition 9.2.3

When the operator L is given by the relation (9.13), the corresponding differential equation (9.1), i.e., the equation

$$\{p(x)\, y'(x)\}' + \{\lambda - q(x)\}\, y(x) = 0. \tag{9.16}$$

is called the '*Sturm-Liouville equation*', and the associated eigenvalue problem is called the '*Sturm-Liouville problem*'.

We shall now present two important theorems associated with self- adjoint eigenvalue problems, in general.

Theorem 9.2.1

The eigenvalues of a self-adjoint eigenvalue problem are real.

Proof

Putting $f(x) = g(x) = \psi(x)$ in equation (9.12), where $\psi(x)$ is an eigenfunction corresponding to the eigen-value λ_0 i.e., where

$$L\,\psi(x) = \lambda_0\, \psi(x), \qquad (L = \text{the given self-adjoint operator}) \tag{9.17}$$

we get,

$$\lambda_0 \int_a^b |\psi(x)|^2\, dx = \overline{\lambda}_0 \int_a^b |\psi(x)|^2\, dx,$$

i.e., $$\lambda_0 = \bar{\lambda}_0, \tag{9.18}$$

and this proves the theorem.

Theorem 9.2.2

For a self-adjoint eigenvalue problem, the two eigenfunctions $\psi_1(x)$ and $\psi_2(x)$, corresponding to two different eigenvalues λ_1 and λ_2, respectively, are orthogonal to each other,

i.e., $$\int_a^b \psi_1(x)\,\overline{\psi_2(x)} = 0 \int_a^b \overline{\psi_1(x)}\,\psi_2(x)\,dx.$$

Proof

We have that

and $$\left.\begin{aligned} L\psi_1 &= \lambda_1\psi_1 \\ L\psi_2 &= \lambda_2\psi_2 \end{aligned}\right\}, \tag{9.19}$$

where L is the given self- adjoint operator.

Then putting $f = \psi_1$ and $g = \psi_2$ in equation (9.12) and using the fact that λ_1 and λ_2 are real, we obtain

$$\lambda_1 \int_a^b \psi_1\,\overline{\psi}_2\,dx = \lambda_2 \int_a^b \psi_1\,\overline{\psi}_2\,dx$$

$$\Rightarrow \qquad \int_a^b \psi_1\,\overline{\psi}_2\,dx = 0 \quad (\text{if } \lambda_1 \neq \lambda_2).$$ Hence the theorem.

The important use of the ideas that have been explained above lies in the following interesting theorem, which we shall just state here, without proof.

Theorem 9.2.3

For a Sturm-Liouville problem described by the O.D.E. (9.16), for which $p(x) > 0$ in $[a, b]$ with any one of the sets of boundary conditions (9.14) and (9.15), there exists a sequence of real eigenvalues λ_n $(n = 0, 1, 2, \ldots)$ such that $\lambda_0 \le \lambda_1 \le \lambda_2 \le \ldots$, and $\lambda_n \to \infty$, as $n \to \infty$.

It is easily checked that the theorem holds good for the particular problem associated with the O.D.E. (9.4) and the conditions (9.5), considered earlier. Several other problems for which the above theorem holds good can be constructed easily, and we have listed some problems at the end of this chapter for this purpose.

It can also be proved (see Eastham [4] for details) *that any function $f(x)$ which satisfies suitable conditions, in an interval $[a, b]$, can be written in the form*

$$f(x) = \sum_{n=0}^{\infty} c_n \psi_n(x), \tag{9.20}$$

where

$$c_n = \int_a^b f(t)\, \overline{\psi_n(t)}\, dt, \tag{9.21}$$

and the $\psi_n(x)$ form an orthonornal set of eigenfunctions.

We shall conclude this chapter by simply making the statement that it can be proved rigorously that *eigen-values do really exist for the Sturm-Liouville operator as given by equation* (9.16), and advise the reader to see Eastham's book for further details.

PROBLEMS

1. Consider the eigenvalue problem

   $$y''(x) + \lambda y(x) = 0, \qquad (0 \le x \le 1)$$

   with

   $$y(1) = \alpha\, y(0), \quad y'(1) = -\,\alpha\, y'(0), \quad (\alpha, \text{ real})$$

   Prove that there are no eigenvalues if $\alpha^2 \neq 1$, and all the eigenvalues are complex if $\alpha^2 = 1$.

2. Show that the non-zero eigenvalues of the eigenvalue problem

   $$y''(x) + \lambda\, y(x) = 0 \qquad (0 \le x \le 1)$$

   with
   $y(0) = 0 = y(1) - by'(1)$, are the solutions of the transcendental equation.

   $$\tan\sqrt{\lambda} = b\,\sqrt{\lambda}, \qquad (b \text{ real}).$$

3. Find the eigenvalues and the corresponding eigen-functions for the problem

   $$y''(x) + \lambda\, y(x) = 0, \qquad (a \le x \le b)$$

   with $y'(a) = y'(b) = 0$.

4. Verify the truth of the theorems 9.2.1-9.2.3 for the case of the eigenvalue problem given by problem 3, above.

CHAPTER - TEN

The Green's Function Technique

In this chapter, we shall describe a very powerful technique, of solving problems involving linear ordinary differential equations, known as the "Green's function Technique". The treatment of this technique, in the present book, is a formal one and, for a more rigorous approach, the reader is advised to take up a study of the subject starting with the book by Friedman [1].

The Green's function technique is well-suited to handle a class of boundary value problems associated with linear ordinary differential equations of any order equal to or greater than 2, and in certain circumstances, this technique can also be used to cast even problems associated with non-linear ordinary differential equations into the equivalent problems of solving integral or integrodifferential equations whose solutions can be derived by various methods (see the book by Kanwal [2]).

10.1 THE BASIC IDEA FOR SECOND ORDER EQUATIONS

The basic idea behind the Green's function technique is as follows :

Let us consider the following two-point boundary value problem associated with a second-order linear ordinary differential equation :

Problem. Solve the O.D.E.

$$Ly = f(x),\ a \le x \le b, \tag{10.1}$$

under the boundary conditions that

$$c_1\, y\,(a) + c_2\, y'\,(a) = 0 \text{ and } d_1\, y\,(b) + d_2\, y'\,(b) = 0, \tag{10.2}$$

with c_1, c_2, d_1, d_2 as certain known constants, where L represents a second-order linear ordinary differential operator of the form

$$\frac{d}{dx}\left(p\,(x)\,\frac{d}{dx}\right) + q\,(x), \tag{10.3}$$

in which the functions p and q represent certain differentiable functions of x in the domain $a \le x \le b$, as is also the case with the inhomogeneous term $f\,(x)$ appearing in the given equation (10.1). [Note: The boundary conditions, of the form (10.2), in each of which only one of the end-points a or b occur are known as "unmixed homogeneous boundary conditions" which are in contrast with the "mixed boundary conditions" for the same problem in which there occur the two end points in a mixed manner].

In order to solve the operator equation (10.1) under the conditions (10.2) it is natural to determine the inverse operator L^{-1} in such a manner that we ultimately may be able to achieve the identity operator I having the property that

$$L^{-1} L = LL^{-1} = I, \tag{10.4}$$

and that eventually we may be able to arrive at the solution of equation (10.1) in the form

$$U = L^{-1} f. \tag{10.5}$$

Because we are interested, in the present context, in the inverse operator L^{-1} which solves a differential equation, it is desirable that we may be able to cast this operator L^{-1} in the form of an integral as given by the relation

$$L^{-1} y = \int_a^b f(x, t)\, y(t)\, dt \tag{10.6}$$

where $g(x, t)$ represents the "Kernel" of the integral operator thus defined, and such a Kernel-function $g(x, t)$ will be called the "Green's function" associated with the differential operator L occurring here, in the case of the above problem, and later on also for other such differential operators appearing in our discussion.

By formal manipulations, we find from the relation (10.6), after using the property (10.4), that

$$\begin{aligned} y(x) \equiv Iy = LL^{-1} y &= L \int_a^b f(x, t)\, y(t)\, dt \\ &= \int_a^b [Lf(x, t)]\, y(t)\, dt \end{aligned} \tag{10.7}$$

where we have pushed the differential operator L inside the sign of integration, in a rather formal manner, by taking into consideration the fact that L is an operator which involves derivatives with respect to x, so that the integration variable t in equation (10.7), can be treated as a parameter.

Writing

$$Lg(x, t) = \delta(x, t) \tag{10.8}$$

we thus obtain, from equation (10.7), that

$$y(x) = \int_a^b \delta(x, t)\, y(t)\, dt, \qquad (a < x < b). \tag{10.9}$$

The above relation (10.9) is the most fascinating identity of Mathematics in the sense that one cannot imagine the existence of an *ordinary* function

$\delta(x, t)$ which, when multiplied by the function $y(t)$ and integrated, gives rise to the same function $y(x)$. It is because of this pecularity of the nature of the function $\delta(x, t)$ that we will regard this "non-ordinary" function as a "generalized" function and treat it to be equal to what is known as the "Dirac's delta function", i.e.,

$$\delta(x, t) = \delta(x - t), \tag{10.10}$$

where the "Dirac's delta function" $\delta(x)$ is defined by means of the following statement:

Definition 10.1

The generalised function, known as the δ function (in short), and written as $\delta(x)$, is zero for every value of x except $x = 0$, where it is undefined and is such that

$$\text{(i)} \quad \int_{-\infty}^{\infty} \delta(x)\, dx = 1,$$

and

$$\text{(ii)} \quad \int_{-\infty}^{\infty} f(x)\, \delta(x)\, dx = f(0), \tag{10.11}$$

where $f(x)$ is a continuously differentiable ordinary function of x.

For a rigorous mathematical definition and various important properties of "Generalized functions", the reader is referred to the book of Lighthill [3].

From what we have described above, it becomes clear then, that if the inverse operator L^{-1} is determined by way of obtaining the Green's function $g(x, t)$ of the operator L in question, which satisfies the equation

$$Lg(x, t) = \delta(x - t), \tag{10.12}$$

for a fixed $t \in (a, b)$, we can solve the given O.D.E. (10.1) by means of the formula

$$y(x) \equiv Iy = L^{-1} Ly = L^{-1} f(x) = \int_a^b g(x, t) f(t)\, dt. \tag{10.13}$$

For the complete specification of the Green's function $g(x, t)$, we must also see that it is such a function which helps satisfy the prescribed boundary conditions on the unknown function $y(x)$ that we wish to determine. We, therefore, take up the unmixed boundary conditions (10.2) prescribed for our problem under discussion here, whilst the method explained is applicable to more general linear differential operators under more general mixed boundary conditions, as will be shown through the examples taken up later on.

The function $y(x)$ as given by the relation (10.13) will satisfy the boundary condition (10.2), if we choose the Green's function $g(x, t)$ satisfying the same conditions as those in equations (10.2), treating t as a constant, i.e., if

$$c_1 g(a, t) + c_2 g'(a, t) = 0, \text{ and } d_1 g(b, t) + d_2 g'(b, t) = 0 \quad (10.14)$$

where 'dash' denotes differentiation w.r.t. x.

We now have a set of relations to determine the Green's function $g(x, t)$ of the operator L in question, and these are supplied by the simplified ODE (10.2) and the boundary conditions (10.14). [We regard equation (10.12) as a simplified ODE, in contrast with the original ODE (10.1) with a possibly complicated right-hand side, in the sense that the right-hand-side of equation (10.12) is zero except when $x = t$.]. The full problem of determining the Green's function $g(x, t)$ can be successfully handled only when we prescribe two more conditions (in the case of the second order equation considered above), because we shall have to determine, altogether, four arbitrary constants, involved in the solution of the two homogeneous ODE of second order that are actually present in the single equation (10.12), because of the fact that $\delta(x - t) = 0$ for $x < t$ and $x > t$, which demands that the two homogeneous equations, that are present in equation (10.12), are :

$$\left.\begin{aligned} Lg(x, t) &= 0, \quad \text{for } x < t \\ \text{and} \qquad Lg(x, t) &= 0, \quad \text{for } x > t \end{aligned}\right\}. \quad (10.15)$$

The two extra conditions that we have talked above are supplied by the following two requirements.

(i) The Green's function $g(x, t)$ is continuous in x for all t,

(ii) The "jump" $[g'(x, t)]$ at $x = t$ of the first derivative (w.r.t. x) of the Green's function is a prescribed function for the second order operator L considered above. In fact, if our operator L is the one as given by the relation (10.3), we first observe that

$$\int_{t-\varepsilon}^{t+\varepsilon} Lg(x, t)\, dx = \int_{t-\varepsilon}^{t+\varepsilon} \delta(x - t)\, dx = 1, \qquad \text{[using (10.11)]}$$

for all $\varepsilon > 0$, which eventually leads to the identity that

$$\left[p(x)\frac{dg}{dx}\right]_{x=t-\varepsilon}^{x=t-\varepsilon} + \int_{t-\varepsilon}^{t+\varepsilon} q(x)\, g(x, t)\, dx = 1,$$

and this, in limit when $\varepsilon - 0$, produces the relation that

$$[g'(x, t)]_{at\, x=t} = 1/p(t), \quad (10.16)$$

because of the fact that p and q represent continuous functions of x, assuming $p(x) \neq 0$, for $x \in (a, b)$.

The determination of the Green's function $g(x, t)$ can thus be completed successfully by solving equations (10.15) satisfying the boundary conditions (10.14), along with the continuity condition (*i*) and the "jump" condition (10.16) whenever the usual consistency and the solvability conditions are satisfied by the resulting system of algebraic equations. The solution of the original boundary value problem posed by equation (10.1) and the conditions (10.2) can finally be determined by utilizing this Green's function $g(x, t)$, in the representation (10.13).

In order to clarify the situations and the technique described above we shall now take up some examples.

Example 10.1. Solve

$$\frac{d^2y}{dx^2} = f(x), \text{ in } 0 \le x \le 1,$$

with

$y(0) = y(1) = 0.$ (unmixed homogeneous boundary conditions)

Solution

The Green's function $g(x, t)$ satisfies the two equations

$$\frac{d^2 g}{dx^2} = \begin{cases} 0, & x < t \\ 0, & x > t \end{cases} \tag{10.17}$$

which can be solved to give

$$g(x, t) = \begin{cases} C_1 x + C_2, & (x < t) \\ D_1 x + D_2, & (x > t) \end{cases} \tag{10.18}$$

where C_1, C_2, D_1 and D_2 are four arbitrary constants.

The boundary conditions for g are : [see (10.14)]

$$g(0, t) = g(1, t) = 0, \tag{10.19}$$

which give

$$C_2 = 0, \; D_1 + D_2 = 0. \tag{10.20}$$

The "continuity" of $g(x, t)$ at $x = t$, gives

$$C_1 t + C_2 = D_1 t + D_2. \tag{10.21}$$

Finally the "jump" condition (10.16) gives that (since $p(x) = 1$)

$$D_1 - C_1 = 1. \tag{10.22}$$

Four equations (10.20) – (10.22) for the four constants C_1, C_2, D_1 and D_2 are easily solved to give

$$C_2 = 0, \; C_1 = t - 1, \; D_1 = t = -D_2 \tag{10.23}$$

and we determine the Green's function $g(x, t)$ associated with the problem as given by the relations

$$g(x, t) = \begin{cases} x(t-1), & x \le t \\ t(x-1), & x \ge t \end{cases} \tag{10.24}$$

The solution of the boundary value problem in question is then given by the relation

$$y(x) = \int_0^1 g(x, t) f(t)\, dt, \tag{10.25}$$

where g is as given by the relations (10.24).

[*Note :* An observation on the Green's function given by the relations (10.24) shows that $g(x, t)$ is a "symmetric" function of x and t, i.e., $g(x, t) = g(t, x)$. This "symmetry" property of the Green's function holds good for the general second-order linear differential operator L that we have been discussing here so far which can be proved without much of a difficulty].

Problems involving Green's functions can also be stated in a slightly different language as in the following example:

Example 10.2. Obtain the Green's function for the operator $L = -\dfrac{d^2}{dx^2} - 4$, involving the boundary conditions $y(0) = y'(1) = 0$.

Solution

The Green's function $g(x, t)$ satisfies the equation

$$\frac{d^2 g}{dx^2} + 4g = -\delta(x - t) = \begin{cases} 0, & x < t \\ 0, & x > t \end{cases} \tag{10.26}$$

and the boundary conditions

$$g(0, t) = g'(1, t) = 0. \tag{10.27}$$

The general solutions of the two equations in (10.26) can be expressed in terms of four arbitrary constants C_1, C_2, C_3 and C_4, in the forms:

$$g(x, t) = \begin{cases} C_1 \sin 2x + C_2 \cos 2x, & (x < t) \\ C_3 \sin 2x + C_4 \cos 2x, & (x > t) \end{cases} \tag{10.28}$$

On using the boundary conditions (10.27) we find that we must have

$$C_2 = 0 \text{ and } C_3 \cos 2 - C_4 \sin 2 = 0. \tag{10.29}$$

We thus obtain that

$$g(x, t) = \begin{cases} C_1 \sin 2x, & (x < t) \\ \dfrac{C_4}{\cos 2} \cos 2(1 - x), & (x > t) \end{cases} \tag{10.30}$$

The continuity condition, of $g(x, t)$ at $x = t$, gives that

$$C_1 \sin 2t = \frac{C_4}{\cos 2} \cos 2(1 - t). \tag{10.31}$$

Next, the "jump" – condition

$$[g'(x, t)]_{x=t} = \frac{1}{p} = -1 \text{ (since } p = -1, \text{ here)}, \tag{10.32}$$

gives

$$C_4 = \frac{1}{2} \sin 2t \tag{10.33}$$

and this, along with the relation (10.31) determines all the constants. We finally arrive at the following expressions for the Green's function :

$$g(x, t) = \begin{cases} \dfrac{1}{2 \cos 2} \sin 2x \cos 2(1-t), & \text{for } x \le t \\ \dfrac{1}{2 \cos 2} \sin 2t \cos 2(1-x), & \text{for } x \ge t. \end{cases} \tag{10.34}$$

We again observe, as expected, that $g(x, t)$ represents a symmetric function of the variables x and t.

We shall now make certain important observations. We find that the Green's function technique as explained above, for second order linear ordinary differential equations can also be successfully utilized to cast certain non-linear second order ordinary differential equations into equivalent integral equations, as in the following example :

Example 10.3. Cast the differential equation

$$\frac{d^2 \phi}{dx^2} + w^2 \phi = F(x, \phi(x)), \quad (0 \le x \le a) \tag{10.35}$$

obeying the end-conditions $\phi(0) = \phi(a) = 0$, into an integral equation, in which F is a differentiable function of its arguments.

Solution

If we assume, for our purposes, that the right-hand side F of the given equation is known, then, from our above discussions on the Green's function technique for second-order equations, we observe that we can write the solution of the problem in the form

$$\phi(x) = \int_0^a g(x, t)\, F(t, \phi(t))\, dt, \tag{10.36}$$

where the Green's function $g(x, t)$ can be easily determined as in *example 10.2* above, in the form

$$g(x, t) = \begin{cases} -(\omega \sin \omega a)^{-1} \sin \omega k \sin \omega (a-t), & (x \le t) \\ -(\omega \sin \omega a)^{-1} \sin \omega (a-x) \sin \omega t, & (x \ge t). \end{cases} \tag{10.37}$$

It is important to note that the unknown function ϕ now appears in equation (10.36) under an integral sign and it is for this reason that this new equation (10.36) is regarded as an "integral equation" for the determination of the function $\phi(x)$. Equation (10.36), incidentally, represents the desired integral equation into which we wanted to recast our problem involving the solution of the differential equation (10.35).

Examples and utility of this type of recasting a given problem of differential equation into an equivalent integral equation are quite interesting and we mention some of them in the exercise at the end of this chapter (see R.P. Kanwal [2] also).

Example 10.4. Cast the boundary value problem

$$\frac{d^2 y}{dx^2} + \lambda y = 0, \; y(0) = 0, \; y'(1) + cy(1) = 0,$$

into an integral equation, for constant values of c and λ.

Solution

Transferring the term λy into the right hand side, the new equation to be solved is

$$\frac{d^2 y}{dx^2} = -\lambda y, \tag{10.38}$$

and, this equation can be attacked for solution by assuming the right-hand side to be a known function, by the aid of the Green's function technique described above. We find that the Green's function $g(x, t)$ is given by

$$g(x, t) = \begin{cases} [1 + C(1-t)\,x]/(1+C), & x \le t \\ [1 + C(1-x)\,t]/(1+C), & x \ge t \end{cases} \tag{10.39}$$

and the solution of (10.38) under the given boundary conditions can be expressed in the form :

$$y(x) = -\lambda \int_0^t g(x, t)\, y(t)\, dt, \tag{10.40}$$

which is an integral equation.

Example 10.5. Cast the Bessel equation

$$x^2 \frac{d^2 y}{dx^2} + x\frac{dy}{dx} + (n^2 x^2 - 1)\, y = 0$$

with end conditions $y(0) = 0 = y(1)$. into an integral equation.

Solution

Rewrite the Bessel equation in the form

$$x^2 \frac{d^2 y}{dx^2} + x\frac{dy}{dx} - y = -n^2 x^2 y. \tag{10.41}$$

If ve then determine the Green's function $g(x, t)$ satisfying

$$x^2 \frac{d^2 g}{dx^2} + x\frac{dg}{dx} - g = \delta(x - t), \tag{10.42}$$

for a fixed $t \in (0, 1)$, with the end conditions

$$g(0, t) = g(1, t) = 0, \tag{10.43}$$

we can solve equation (10.41), with known right-hand-side, in the form

$$y(x) = -n^2 \int_0^1 t^2 g(x, t) y(t) dt, \tag{10.44}$$

which is an integral equation for the determination of $y(x)$. We find that the homogeneous equation (10.42) is of the Euler-Cauchy type (see Chapter 2) and, then, the Green's function can be easily determined in the form

$$g(x, t) = \begin{cases} -\dfrac{x}{2t}(1 - t^2), & x \leq t \\ -\dfrac{t}{2x}(1 - x^2), & x \geq t. \end{cases} \tag{10.45}$$

The following general theorem involving the Green's function for the O.D.E. of the type

$$L(w) \equiv \frac{d}{dx}\left(p(x)\frac{dw}{dx}\right) + q(x)w = 0 \qquad (a \leq x \leq b)$$

can be proved easily by the principles described above, and, then, the calculations of the Green's functions of several linear differential operators of second order can be carried out by just using this theorem.

Theorem 10.1

Let $w_1(x)$ be a solution of $L(w) = 0$, satisfying the unmixed homogeneous boundary condition $B_1(w) \equiv C_1 w(a) + C_2 w'(a) = 0$, and let $w_2(x)$ be a solution of the same equation satisfying a different unmixed boundary condition $B_2(w) \equiv d_1 w(b) + d_2 w'(b) = 0$, where C_1, C_2, d_1, d_2 are constants. Then the Green's function for L with the boundary conditions B_1 and B_2 is given by the formula:

$$G(x, t) = +\frac{w_1(x) w_2(t) H(t - x) + w_2(x) w_1(t) H(x - t)}{p(t) J(w_2, w_1)}, \tag{10.46}$$

where $J(w_2, w_1)$ represents the Wronskian determinant as given by

$$J(w_2, w_1) = w'_2 w_1 - w'_1 w_2 \tag{10.47}$$

evaluated at $x = t$, and $H(x)$ represents the Heaviside's step-function defined as

$$H(x) = \begin{cases} 0, & x < 0 \\ 1, & x > 0 \end{cases} \tag{10.48}$$

In order to show the utility of the formula (10.46), we shall take up the problem in example 10.5 considered above, very briefly.

The operator L involved in equation (10.42) is

$$L = x^2 \frac{d^2}{dx^2} + x \frac{d}{dx} - 1$$

and the two solutions can be taken as $w_1(x) = x$, $w_2 = \dfrac{1 - x^2}{x}$.

This gives the Wronskian $J(w_1, w_2)$, evaluated at t, in the form :

$$J(w_1, w_2)\,|_{x=t} = \begin{vmatrix} x & 1 - x^2/x \\ 1 & -\dfrac{(1 + x^2)}{x^2} \end{vmatrix}_{x=t} = -\frac{2}{t};$$

and, then, the Green's function in example 10.5 above, is given by the formula (see Theorem 10.1):

$$g(x, t) = \begin{cases} -x(1 - t^2)/2t, & x \le t \\ -t(1 - x^2)/2x, & x \ge t \end{cases}$$

which is the same as that given by the relations (10.45), derived earlier.

Whilst in all the discussions so far, the ordinary differential equation considered has the form

$$LY \equiv \frac{d}{dx}\left[p(x)\frac{dy}{dx}\right] + q(x)\,y = 0, \qquad (10.49)$$

any general second order linear ordinary differential equation

$$a(x)\frac{d^2y}{dx^2} + b(x)\frac{dy}{dx} + c(x)\,y = 0, \qquad (10.50)$$

where a, b, c are continuously differentiable, and a does not vanish anywhere in the interval of x under consideration, can also be cast into the form (10.49) in the following manner : Multiply equation (10.50) by the expression

$$M = \left\{\frac{1}{a(x)} \exp\left(\int \frac{b(x)}{a(x)}\,dx\right)\right\}$$

whenever this expression is meaningful, and rewrite equation (10.50) in the form:

$$\frac{d}{dx}\left[\exp\left(\int \frac{b(x)}{a(x)}\,dx\right)\frac{dy}{dx}\right] + \frac{c(x)}{a(x)} \exp\left(\int \frac{b(x)}{a(x)}\,dx\right) y = 0, \qquad (10.51)$$

and this is of the type (10.49), with

$$p(x) = \exp\left(\int \frac{b(x)}{a(x)}\,dx\right)$$

and $$q(x) = \frac{c(x)}{a(x)} \exp\left(\int \frac{b(x)}{a(x)}\,dx\right) \qquad (10.52)$$

In order to exactly understand the reason behind choosing the particular expression M, we set $y(x) = u(x)\,v(x)$, where u and v are twice differentiable functions of x, so that we may express equation (10.49) as

$$\frac{d}{dx}\left[(av)\frac{du}{dx}\right] + \left[a\frac{dv}{dx} - v\frac{da}{dx} + bv\right]\frac{du}{dx} + \left[a\frac{d^2v}{dx^2} + b\frac{dv}{dx} + cv\right]u = 0. \tag{10.53}$$

If we then allow the function v to be so chosen that

$$a\frac{dv}{dx} - \left(\frac{da}{dx} - b\right)v = 0, \tag{10.54}$$

so that the coefficient of $\frac{du}{dx}$ in the left of equation (10.53) vanishes, we obtain an equation in u which is of the type (10.49). Now, an appropriate choice of the function v is supplied by a solution of the first order O.D.E. (10.54), and this can be taken as

$$v(x) = \exp\left[\int\left(\frac{\frac{da}{dx} - b}{a}\right)dx\right] = a(x)\exp\left(-\int\frac{b(x)}{a(x)}dx\right) \tag{10.55}$$

It is then easy to understand the choice of M in the approach described above which helps recasting equations of the general form (10.50) into the form (10.49) and once such a recasting is done, the theory described above, for the determination of the Green's function, can be applied in a straightforward manner, with slight adjustments.

Example 10.6. Determine the Green's function of the operator

$$L = \frac{d^2}{dx^2} + 4\frac{d}{dx} + 4,$$

with $y(0) = y(1) = 0$.

Solution

The equation $Ly = 0$ can be recast after multiplying by e^{4x} into the form (cf. equation (10.51))

$$\frac{d}{dx}\left[\exp\left(\int 4dx\right)\frac{dy}{dx}\right] + 4\exp\left(\int 4dx\right)y = 0$$

i.e.,

$$\frac{d}{dx}\left[e^{4x}\frac{dy}{dx}\right] + 4e^{4x}y = 0, \tag{10.56}$$

giving $p(x) = e^{4x}$ and $q(x) = 4e^{4x}$, for this example.

Then, using Theorem 10.1, the Green's function $g(x, t)$ can be easily computed, after observing that the two linearly independent solutions of the homogeneous equation $Ly = 0$ are given by

$$y_1(x) = e^{-2x} \text{ and } y_2(x) = xe^{-2x}. \tag{10.57}$$

Further details are left to the reader. The ultimate result is that the Green's function $g(x, t)$ for the above operator is obtained in the form:

$$g(x, t) = [x(t-1) H(t-x) + t(x-1) H(x-t)] e^{-2(x+t)}, \quad (10.58)$$

where $H(x)$ is the Heaviside function introduced earlier. We find that $g(x, t)$ is a symmetric function of x and t.

10.2 HIGHER ORDER EQUATIONS

Boundary value problems involving a linear ordinary differential equation of order n (> 2), with prescribed end conditions can also be handled successfully, by extending the ideas for the Green's function for second order equations explained in the previous section, in a straightforward manner and, we shall consider below two typical problems to clarify the situations.

Problem 1. Solve the boundary value problem

$$\frac{d^4 y}{dx^4} = + f(x),$$

under the conditions $y(0) = 0 = y'(0)$, $y(1) = 0 = y'(1)$.

Solution

We seek the Green's function $G(x, t)$ defined for x, t (such that $0 \le x, t \le 1$), which satisfies the following equation and conditions :

(i) $\dfrac{d^4 G}{dx^4} = \delta(x-t), 0 < x, t < 1,$

(ii) $G(0, t) = 0$

(iii) $\dfrac{dG}{dx}\Big|_{x=0} = 0$

(iv) $G(1, t) = 0$

(v) $\dfrac{dG}{dx}\Big|_{x=1} = 0,$

(vi) $G, \dfrac{dG}{dx}, \dfrac{d^2 G}{dx^2}$ are continuous for all $x, t \in (0, 1)$,

(vii) $\left[\dfrac{d^3 G}{dx^3}\right]_{x=t} \equiv \left(\dfrac{d^3 G}{dx^3}\right)_{x=t+0} - \left(\dfrac{d^3 G}{dx^3}\right)_{x=t-0} = 1$

(obtained by integrating (i) w.r.t. x between $-\varepsilon$ and $t+\varepsilon$, and taking limit as $\varepsilon \to 0$, as was done earlier, for second order equations.

The solutions of the equation (i) will involve 4 plus 4 (= 8) arbitrary constants altogether, and their explicit forms are as given below.

$$G(x, t) = \begin{cases} C_0 + C_1 x + C_2 x^2 + C_3 x^3, & x \le t \\ D_0 + D_1 x + D_2 x^2 + D_3 x^3, & x \ge t. \end{cases} \quad (10.59)$$

In order to determine these eight constants which can be regarded as dependent on the other variable t, we have to utilize eight conditions and, these conditions are supplied above in the forms (ii) to (vii).

When the algebraic manipulations are completed, we find the Green's function $G(x, t)$, ultimately in the following form:

$$G(x, t) = \begin{cases} \frac{1}{6} x^2 (1-t)^2 (2xt + x - 3t), & x \le t \\ \frac{1}{6} t^2 (1-x)^2 (2xt + t - 3x), & x \ge t. \end{cases} \tag{10.60}$$

The solution of the boundary value problem in question can then be determined by using the formula

$$y(x) = \int_0^1 G(x, t) f(t)\, dt, \tag{10.61}$$

as before (cf. the relation (10.13)).

Problem 2. Reduce the boundary value problem

$$\frac{d^2}{dx^2}\left\{(4+x)^3 \frac{d^2 y}{dx^2}\right\} - \lambda (4+x) y = 0, \qquad (\lambda = \text{a constant})$$

$y(0) = 0 = y(1)$, $y''(0) = 0 = y'''(1)$, to an integral equation.

Solution

We first rewrite the given O.D.E. in the form

$$\frac{d^2}{dx^2}\left\{(4+x)^3 \frac{d^2 y}{dx^2}\right\} = \lambda (4+x) y = \lambda u(x) \quad \text{(say)}. \tag{10.62}$$

Assuming the right hand side of equation (10.62) to be a known function, we can express the solution of this equation, under the given boundary conditions $y(0) = 0 = y(1)$ and $y''(0) = 0 = y'''(1)$, in the integral form

$$y(x) = \lambda \int_0^1 G(x, t)\, u(t)\, dt \equiv \lambda \int_0^1 (4+t)\, G(x, t)\, y(t)\, dt, \tag{10.63}$$

which represents the desired integral equation, where the Green's function $G(x, t)$ satisfies the following equations and conditions.

(i) $\frac{d^2}{dx^2}\left\{(4+x)^3 \frac{d^2 G}{dx^2}\right\} = \delta(x-t)$,

(ii) $G(0, t) = 0$,

(iii) $G(1, t) = 0$,

(iv) $\frac{d^2 G}{dx^2}\Big|_{x=0} = 0$,

(v) $\frac{d^3 G}{dx^3}\Big|_{x=1} = 0$,

(vi) $G, \frac{dG}{dx}, \frac{d^2 G}{dx^2}$ are continuous for all $x, t \in (0, 1)$, and

(vii) The jump $\left[\dfrac{d}{dx}\left[(4+x)^3\dfrac{d^2G}{dx^2}\right]\right]_{x=t} = 1,$

(obtained by integrating (i) w.r.t. x, between $t-\varepsilon$ and $t+\varepsilon$, and taking limit as $\varepsilon \to 0$).

The solution of equation (i) may be expressed in the form

$$G(x,t) = \begin{cases} \dfrac{C_0}{4+x} + C_4 \ln(4+x) + C_2 x + C_3, & x \le t \\ \dfrac{D_0}{4+x} + D_1 \ln(4+x) + D_2(x-1) + D_3, & x \ge t \end{cases} \tag{10.64}$$

where the eight constants C_0, C_1, C_2, C_3 and D_0, D_1, D_2, D_3, which may depend on the other variable t, will have to be determined by using the eight conditions provided by the relations (ii) to (vii).

We find that the following relations are obtained for these constants :

$$C_0 = 2 + 3t,\ D_0 = \frac{5}{2}t,\ C_2 = 1 + \frac{3}{2}t,\ D_1 = \frac{3}{2}t,$$

$$C_3 = -\left(\frac{C_0}{4} + C_1 \ln 4\right),\ D_3 = -\left(\frac{D_0}{5} + D_1 \ln 5\right),$$

$$D_2 = D_3 - C_3 - \frac{2}{(4+t)} - \ln(4+t),$$

$$C_2 = D_2 - \frac{1}{2(4+t)}.$$

Using these relations (10.65), in equations (10.64), we obtain the rather complicated looking Green's function $G(x, t)$ for the problem at hand and, finally, the reduction of the problem to an integral equation can be completed by using the result (10.63).

It should be noted that the Green's function for the above problem is not a symmetric function of x and t.

10.3 INITIAL VALUE PROBLEMS

A technique allied to the Green's function technique described above can be utilized to solve certain initial value problems. In order to solve the initial value problem of the type

$$L(y) \equiv \frac{d}{dx}\left(p(x)\frac{dy}{dx}\right) + q(x)\,y = F(x), \quad (a < x < b) \tag{10.66}$$

with the homogeneous initial conditions

$$y(a) = y'(a) = 0, \tag{10.67}$$

we express the solution in the form

$$y(x) = \int_a^x G(x, t) F(t)\, dt, \tag{10.68}$$

where the Green's function $G(x, t)$ can be determined as follows.

Differentiating the relation (10.67) w.r.t. x, assuming the differentiability of G, we obtain

$$y'(x) \equiv \frac{dy}{dx} = \int_a^x \frac{\partial G}{\partial x} F(t)\, dt + G(x, x) \quad F(x), \tag{10.69}$$

and

$$y''(x) \equiv \frac{d^2y}{dx^2} = \int_a^x \frac{\partial^2 G}{\partial x^2} F(t)\, dt + \left(\frac{\partial G}{\partial x}\right)_{t=x} F(x) + \frac{\partial}{\partial x}[G(x, x) F(x)]. \tag{10.70}$$

Then, assuming that we have chosen $G(x, t)$ satisfying the relation

$$G(x, x) = 0, \tag{10.71}$$

we easily derive that

$$L(y) = \int_a^x \left[\frac{d}{dx}\left(p(x)\frac{\partial G}{\partial x}\right) + q(x) G(x, t)\right] F(t)\, dt + p(x)\left(\frac{\partial G}{\partial x}\right)_{t=x} F(x). \tag{10.72}$$

We, thus, observe from the above results (10.69) to (10.72), that the relation (10.68) will solve the initial value problem completely, if the Green's function $G(x, t)$ satisfies the following initial value problem :

$$\frac{d}{dx}\left(p(x)\frac{\partial G}{\partial x}\right) + q(x) G = 0, \tag{10.73}$$

with the initial conditions :

$$G(x, t)\,|_{x=t} = 0 \text{ and } \left(\frac{\partial G}{\partial x}\right)_{x=t} = \frac{1}{p(t)}. \tag{10.74}$$

The following example clarifies the situations described above.

Example 10.7. Solve the initial value problem

$$y'' + 4y = \sin 2x, \quad 0 < x < 1,$$

with

$$y(0) = y'(0) = 0.$$

Solution

In terms of the Green's function $G(x, t)$, as described above, the solution can be expressed as

$$y(x) = \int_0^x G(x, t) \sin 2t \; dt, \tag{10.75}$$

where $G(x, t)$ satisfies the following initial value problem :

$$\frac{d^2 G}{dx^2} + 4G = 0, \tag{10.76}$$

with

$$G(t, t) = 0 \text{ and } \left(\frac{\partial G}{\partial x}\right)_{x=t} = 1. \tag{10.77}$$

The solution for the initial value problem for G is easily obtained as

$$G(x, t) = \frac{1}{2} \sin 2(x - t), \tag{10.78}$$

and then, the solution (10.75) is given by

$$y(x) = \frac{1}{2} \int_0^x \sin 2t \sin 2(x - t)\, dt = \frac{1}{8} \sin 2x - \frac{x}{4} \cos 2x. \tag{10.79}$$

The above solution (10.79) can also be derived by the standard methods described in Chapter 2.

The case of inhomogeneous initial conditions replacing the homogeneous ones as given by the relations (10.67), associated with the differential equation (10.66), can be handled simply by modifying the solution (10.68) in the form

$$y(x) = \int_a^x G(x, t) F(t)\, dt + C_1 y_1(x) + C_2 y_2(x), \tag{10.80}$$

where C_1 and C_2 are arbitrary constants and y_1 and y_2 represent two linearly independent solutions of the homogeneous equation (10.66).

PROBLEMS

1. The Functions $u_1(x)$ and $u_2(x)$ are two independent solutions of the equation

$$\frac{d}{dx}\left[p(x) \frac{du}{dx}\right] - q(x) u = 0, \quad (0 \le x \le a),$$

where $p(x)$ is a nonvanishing differentiable function in $(0, a)$ and $q(x)$ is continuous in $(0, a)$.

Show that the Greens function for the boundary conditions $\alpha u + \beta u' = 0$ at $x = 0$ and $\gamma u + \delta u' = 0$, at $x = a$, ($\alpha, \beta, \gamma, \delta$ are constants) can be obtained in the form

$$G(x, t) = \begin{cases} A\, u_1(x)\, u_2(t), & (0 \le x \le t) \\ A\, u_1(t)\, u_2(x), & (t \le x \le a) \end{cases}$$

where $u_1(0) = \beta$, $u'_1(a) = -\alpha$, $u_2(a) = \delta$, $u'_2(a) = -\gamma$, and A is a suitable constant.

2. Determine the Green's function for the operator $L = \dfrac{d^2}{dx^2}$, $(0 \le x \le 1)$, involving the boundary conditions $y(0) = 0$ and $y'(1) + \mu y(1) = 0$, where

μ is a constant, and hence obtain an equivalent integral equation for the problem

$$\frac{d^2 y}{dx^2} + \lambda y = 0,$$

with the same boundary conditions.

3. Obtain equivalent integral equations for each of the following problems :
   (i) $y'' + y = 0,\ y(0) = 0,\ y'(0) = 1$
   (ii) $y'' + xy = 1,\ y(0) = y(1) = 0$
   (iii) $y'' + y = x,\ y'(0) = y'(1) = 0$
   (iv) $y'' + y = x,\ y(0) = 1,\ y'(1) = 0$
   (v) $y'' + (y'/2x) + \lambda x^{1/2}\, y = 0,\ y(0) = y(1) = 0$
   (vi) $y'' - y = x^2,\ y(0) = y(1) = 0,$

4. Determine the Green's function for the Bessel operator $L = \dfrac{d}{dx}\left(x\dfrac{d}{dx}\right) - n^2/x$, $(n \neq 0$, constant) $(0 \leq x \leq 1)$, involving the boundary conditions $y(0) = y(1) = 0$. Hence obtain an equivalent integral equation for the problem
   $x^2 y'' + xy' + (\lambda x^2 - n^2)\, y = 0$, ($\lambda = a$ parameter) with the same boundary conditions.

5. Reduce the initial value problem

$$y'' + \lambda x^2 y = 0, \qquad (\lambda = a \text{ constant})$$

   with $y(0) = 0,\ y'(0) = 1$, to an integral equation.

6. Solve the following boundary value problems
   (i) $x^2 y'' - xy' = x^3 e^x$, with $y(0) = 0,\ y'(1) = 1$
   (ii) $x^2 y'' - (3x^2 + 2x)\, y' + (3x + 2)\, y = -2x^3$, with $y(1) = 0,\ y'(1) = 2$
   (iii) $y''' - 8y = 4e^{2x}$, with $y(0) = y'(0) = 0,\ y(1) = 1$
   (iv) $x^3 y''' + 3x^2 y'' = 1$, with $y(1) = y'(1) = 0,\ y(2) = 1$

7. Solve the following initial value problems
   (i) $y'' - 2y' + 5y = e^{3x}$, with $y(0) = y'(0) = 3$
   (ii) $2y'' + y' - 3y = \cos 2x$, with $y(0) = 0,\ y'(0) = 1$
   (iii) $y'' + 4y' + 13y = x$, with $y(0) = y'(0) = 0.$
   (iv) $3y'' + 2y' + 3y = xe^{-2x}$, with $y(0) = 1,\ y'(0) = -1$

REFERENCES

1. B. Friedman "principles and Techniques of Applied Mathematics". John Wiley and Sons, Inc, New York, 1956.
2. R.P. Kanwal "Linear Integral Equations: Theory and Techniques". Academic Press, New York and London, 1971.
3. M.J. Lighthill, "Introduction to Fourier Analysis and Generalized Functions", Cambridge University Press, 1964.

Appendix

In the Appendix to this book, we shall briefly discuss some of the other special functions, such as the Gamma and the Beta functions which have been utilized in the main text, and the error functions, the Hypergeometric functions, as well as the Chebyshev polynomials whose study is very much related to the subject of interest in the present book.

A.1 THE GAMMA AND THE BETA FUNCTIONS, AND THE CONFLUENT HYPERGEOMETRIC FUNCTIONS

The integral

$$\Gamma(n) = \int_0^\infty e^{-x} x^{n-1}\, dx \tag{A.1}$$

is known to be convergent if $n > 0$ and, in that event, it is also said to define a function of the variable n, known as the *Gamma function*.

If n is zero or a negative integer, we accept that $\Gamma(n) \to \infty$ and if n is any other negative real number, other than a negative integer, we define the Gamma function by means of the relation

$$\Gamma(n) = \frac{1}{n}\Gamma(n+1), \tag{A.2}$$

which also holds good for positive n, as can be easily verified by the relation (A.1).

As a utility of the above definitions (A.1) and (A.2), we can easily prove the following important results:

(i) $\Gamma(1) = 1$,

(ii) $\Gamma(n+1) = n\,!$, if n is a positive integer,

(iii) $\Gamma(1/2) = \sqrt{\pi}$,

(iv) $\Gamma(x)\,\Gamma(1-x) = \pi \operatorname{cosec}(x\,\pi)$, $0 < x < 1$,

(v) $\Gamma(1/2)\,\Gamma(2n) = 2^{2n-1}\,\Gamma(n)\,\Gamma(n+1/2)$,

(vi) $\Gamma(-3/2) = \frac{4}{3}\sqrt{\pi}$,

(vii) $\Gamma(x+1) = \lim\limits_{n \to \infty} [n!\, n^x / \{(x+1)(x+2) \ldots (x+n)\}],\ x > 0,$

The result (vii) also gives that

(viii) $\dfrac{d}{dx}\{\log \Gamma(x+1)\} = \lim\limits_{n \to 0}\left(\log n - \dfrac{1}{x+1} - \dfrac{1}{x+1} \ldots - \dfrac{1}{x+n}\right)$

Allowing x to tend to zero in (viii) we obtain,

(ix) $\gamma = -\left[\dfrac{d}{dx}\log \Gamma(x+1)\right]_{x=0} = -\int_0^\infty e^{-t}\log t\,dt$, by (A.1),

where

$$\gamma = \lim_{n \to \infty}\left(1 + \frac{1}{2} + \frac{1}{3} + \ldots + \frac{1}{n} - \log n\right) \tag{A.3}$$

The constant γ above, is known as *Euler's constant* whose approximate value is 0.5772.

We also define the *Beta function* $B(m, n)$, by using the integral formula,

$$B(m, n) = \int_0^1 x^{m-1}(1-x)^{n-1}\,dx, \tag{A.4}$$

where the integral converges if $m > 0$, $n > 0$.

We can easily deduce the following useful formula:

(a) $B(m, n) = \dfrac{\Gamma(m)\,\Gamma(n)}{\Gamma(m+n)}$,

(b) $B(m, n) = 2\displaystyle\int_0^{\pi/2} (\sin\theta)^{2m-1}(\cos\theta)^{2n-1}\,d\theta.$

The following are some results involving the Beta and the Gamma functions which have been used in our discussion in the main text:

(1) $(\alpha)_n = \alpha(\alpha+1)\ldots(\alpha+n-1) = \Gamma(\alpha+n)/\Gamma(\alpha)$

(2) $(\alpha-1)(\alpha)_{n-1} = (\alpha-1)_n,$

(3) $\dfrac{n!}{(n-s)!} = (-1)^s(-n)_s,\ (n > s)$

(4) $\left(\dfrac{1}{2} - 2m\right)_{m-s} = \dfrac{\Gamma(2m+1/2)(-1)^{m-s}}{\Gamma(m+1/2)(m+1/2)_s}$

whenever all the terms are well-defined.

The INCOMPLETE Gamma and Beta functions are defined by means of the following integrals:

(i) *First kind Incomplete Gamma function*

$$\gamma(\alpha, x) = \int_0^x e^{-t}\,t^{\alpha-1}\,dt,\quad (\alpha > 0,\ x \text{ real})$$

(ii) *Second kind Incomplete Gamma function*

$$\Gamma(\alpha, x) = \int_x^\infty e^{-t}\, t^{\alpha-1}\, dt \equiv \Gamma(\alpha) - \gamma(\alpha, x), \quad (\alpha > 0, x \text{ real})$$

(iii) *Incomplete Beta function*

$$B(x, y, \alpha) = B_\alpha(x, y) = \int_0^\alpha t^{x-1}(1-t)^{y-1}\, dt,$$

$$0 < \alpha < 1, x > 0, y > 0.$$

The following properties of the incomplete Gamma functions can be deduced easily:

(*) $\gamma(\alpha + 1, x) = \alpha\, \gamma(\alpha, x) - x^\alpha e^{-x}$,

(**) $\Gamma(\alpha + 1, x) = \alpha\, \Gamma(\alpha, x) + x^\alpha e^{-x}$,

(***) $\gamma(\alpha, x) = \alpha^{-1} x^\alpha e^{-x}\, \Phi(1, 1 + \alpha; x)$
$= \alpha^{-1} x^\alpha\, \Phi(\alpha, 1 + \alpha; -x)$,

(****) $\dfrac{d\gamma(\alpha, x)}{dx} = -\dfrac{d\Gamma(\alpha, x)}{dx} = x^{\alpha-1} e^{-x}$,

and

(*****) (a) $\dfrac{d^n}{dx^n}[x^{-\alpha} \gamma(\alpha, x)] = (-1)^n x^{-\alpha-n} \gamma(\alpha + n, x)$,

(b) $\dfrac{d^n}{dx^n}[e^x \gamma(\alpha, x)] = (-1)^n (1-\alpha)_n\, e^x \gamma(\alpha - n, x)$,

where

$$\Phi(a, c; x) = \frac{\Gamma(c)}{\Gamma(a)\,\Gamma(c-a)} \int_0^1 e^{xu} u^{a-1} (1-u)^{c-a-1}\, du, \quad (Re(c) > Re(a) > 0)$$

$$= \sum_{r=0}^{\infty} \frac{(\alpha)_r}{(\gamma)_r \cdot r!} x^r, \tag{A.5}$$

which can be proved by expanding e^{xu} in powers of x and integrating term by term.

The function $\Phi(a, c; x)$ is called *the Confluent hyper-geometric function* which is also denoted by the symbol ${}_1F_1(a, c; x)$.

It can be shown directly that the function $\Phi(a, c; x)$ satisfies the O.D.E., known as the 'Confluent hypergeometric equation', as given by

$$x\frac{d^2 y}{dx^2} + (c - x)\frac{dy}{dx} - ay = 0, \tag{A.6}$$

and this is the limiting form of the 'Hypergeometric equation' as given by

$$x(1-x)\frac{d^2 y}{dx^2} + \{c - (1 + a + b)x\}\frac{dy}{dx} - aby = 0, \tag{A.7}$$

as $b \to \infty$, for the equation (A.7) can be cast into the form:

$$\bar{x}\left(1-\frac{\bar{x}}{b}\right)\frac{d^2\bar{y}}{d\bar{x}^2}+\left\{c-\left(1+\frac{a+1}{b}\right)\bar{x}\right\}\frac{d\bar{y}}{d\bar{x}}-a\bar{y}=0. \tag{A.8}$$

$(\bar{x} = bx)$

It can also be proved that the function defined by

$${}_2F_1(a, b; c; x) = \sum_{r=0}^{\infty} \frac{(a)_r (b)_r}{r!\,(c)_r} x^r, \tag{A.9}$$

represents a solution of the Hypergeometric equation (A.7), which is valid in a neighbourhood of the origin ($x = 0$), which is a regular singular point of equation (A.7).

Further properties of the Hypergeometric functions are listed in section A.3.

Some properties of the incomplete Beta functions, $B_\alpha(x, y)$ are given below :

(1) $B_\alpha(x, y) = B_\alpha(y, x) = x^{-1}\alpha^x\,{}_2F_1(x, 1-y; x+1; \alpha)$,

(2) $B(x, y, 1) = B(x, y)$,

(3) If $I(x, y, \alpha) = \dfrac{B(x, y, \alpha)}{B(x, y)}$, then

(i) $I(x, y, \alpha) = 1 - I(y, x, 1-\alpha)$

(ii) $I(x, y, \alpha) = \alpha I(x-1, y, \alpha) + (1-\alpha) I(x, y-1, \alpha)$,

(iii) $(x+y) I(x, y, \alpha) = xI(x+1, y, \alpha) + yI(x, y+1, \alpha)$,

(iv) $(x+y-xy) I(x, y, \alpha) = x(1-\alpha) I(x+1, y-1, \alpha) + yI(x, y+1, \alpha)$.

A.2 THE ERROR FUNCTIONS

For real values of the argument x, we define the 'Error function' Erf (x) and the 'Complementary Error function' Erf c (x), by means of the following convergent integrals:

(i) $\text{Erf}(x) = \int_0^x e^{-t^2}\,dt,$

and

(ii) $\text{Erf}\,c(x) = \int_0^\infty e^{-t^2}\,dt.$

We can easily prove the following results:

(a) $\text{Erf}(x) = \frac{1}{2}\gamma\left(\frac{1}{2}, x^2\right)$

$= x\,\Phi\left(\frac{1}{2}, \frac{3}{2}; -x^2\right)$

$$= xe^{-x^2}\,\phi\left(1, \frac{3}{2}; x^2\right)$$

(b) $\text{Erfc}\,(x) = \frac{1}{2}\sqrt{\pi} - \text{Erf}\,(x) = \frac{1}{2}\Gamma\left(\frac{1}{2}, x^2\right)$

(c) $\text{Erf}\,(x) = e^{-x2} \sum_{n=0}^{\infty} \frac{x^{2n+1}}{(3/2)_n}$,

(d) $\int_0^{\infty} e^{-a^2 t^2 - bt}\,dt = a^{-1}\exp(b^2/4a^2)\,\text{Erf}\,c\,(b/2a) \qquad (a, b > 0)$

(e) $\int_0^{\infty} \text{Erf}\,(at)\,e^{-st}\,dt = s^{-1}(\exp(s^2/4a^2))\,\text{Erf}\,c\,(s/2a) \qquad (a, s > 0)$

(f) $\int_0^{\infty} \text{Erf}\,((at)^{1/2})\,e^{-st}\,dt = \frac{1}{2}(a\,\pi)^{1/2}\,s^{-1}\,(a+s)^{-1/2} \qquad (a, s > 0)$

(g) $\int_0^{\infty} \text{Erf}\,c\,((at^{-1/2}))\,e^{-st}\,dt = \frac{1}{2}\sqrt{\pi}\,s^{-1}\,e^{-2as^{1/2}} \qquad (a, s > 0)$

(h) $\int_0^{1} e^{-a^2 t^2}\frac{dt}{1+t^2} = e^{a^2}\left[\frac{\pi}{4} - (\text{Erf}\,(a))^2\right], \qquad (a > 0)$

and several other useful results can be derived in a straightforward manner (see Lebedev [8], for example).

A.3 THE HYPERGEOMETRIC FUNCTIONS

The Hypergeometric function ${}_2F_1\,(\alpha, \beta; \gamma; x)$ and the Confluent Hypergeometric function ${}_1F_1\,(\alpha, \beta, x)$ have already occurred in section A.1. In this section, we only list some important relations associated with these special functions of very frequent appearance in solutions of problems of Mathematical Physics. We have, for proper choices of the parameters α, β, γ and the variable x that

(i) ${}_2F_1\,(\alpha, \beta; \gamma; 1) = \dfrac{\Gamma(\gamma)\,\Gamma(\gamma-\alpha-\beta)}{\Gamma(\gamma-\alpha)\,\Gamma(\gamma-\beta)}$,

(ii) ${}_2F_1\,(\alpha, \beta; \beta-\alpha+1; -1) = \dfrac{\Gamma(1+\beta-\alpha)\,\Gamma\left(1+\frac{1}{2}\beta\right)}{\Gamma(1+\beta)\,\Gamma\left(1+\frac{1}{2}\beta-\alpha\right)}$,

(iii) ${}_2F_1\,(\alpha, \beta; \gamma; x) = (1-x)^{-\alpha}\,{}_2F_1\left(\alpha, \gamma-\beta; \gamma; \dfrac{x}{x-1}\right)$

(iv) ${}_2F_1\,(\alpha, \beta; \gamma; x) = (1-x)^{-\beta}\,{}_2F_1\left(\gamma-\alpha, \beta; \gamma; \dfrac{x}{x-1}\right)$,

(v) ${}_2F_1\,(\alpha, \beta; \gamma; x) = (1-x)^{\gamma-\alpha-\beta}\,{}_2F_1\,(\gamma-\alpha, \gamma-\beta; \gamma; x)$

(vi) $(\alpha-\beta)\ {}_2F_1(\alpha,\beta;\gamma;x)=\alpha_2 F_1(\alpha+1,\beta;\gamma;x)$
$$-\beta_2 F_1(\alpha,\beta+1;\gamma;x),$$

(vii) ${}_1F_1(\alpha,\gamma;x)=e^x\ {}_1F_1(\gamma-\alpha\,\gamma;-x)$

(viii) $\dfrac{d}{dx}\{{}_1F_1(\alpha,\gamma;x)\}=\dfrac{\alpha}{\gamma}\ {}_1F_1(\alpha+1;\gamma+1;x)$

(ix) $\alpha_1 F_1(\alpha+1,\gamma+1;x)+(\gamma-a)\ {}_1F_1(\alpha,\gamma+1;x)-\gamma_1 F_1(\alpha,\gamma;x)=0$

(x) $(x+\alpha)\ {}_1F_1(\alpha+1,\gamma+1;x)+(\gamma-\alpha)\ {}_1F_1(\alpha,\gamma+1;x)$
$$-\gamma_1 F_1(\alpha+1,\gamma;x)=0$$

(xi) $\alpha_1 F_1(\alpha+1,\gamma;x)+(\gamma-2\alpha-x)\ {}_1F_1(\alpha,\gamma;x)$
$$+(\alpha-\gamma)\ {}_1F_1(\alpha-i,\gamma;x)=0$$

and

(xii) $(\alpha-\gamma)\,x_1 F_1(\alpha,\gamma+1;x)+\gamma(x+\gamma-1)_1 F_1(\alpha,\gamma;x)$
$$+\gamma(\gamma-1)_1 F_1(\alpha,\gamma-1;x)=0.$$

Similar other relations can be found in the book of Sneddon [12].

A.4 THE CHEBYSHEV POLYNOMIALS

The Chebyshev polynomials $T_n(x)$ and $U_n(x)$ of the 'first' and 'second' kinds, respectively are defined through the Hypergeometric functions, by the following relations:

(i) $T_n(x)={}_2F_1\left(-n,n;\dfrac{1}{2};\dfrac{1-x}{2}\right)$

and

(ii) $U_n(x)=(n+1)\ {}_2F_1\left(-n,n+1;\dfrac{3}{2};\dfrac{1-x}{2}\right)$

where n is a positive integer or zero and $|x|<1$.

We easily deduce, by setting $x=\cos\theta,\ -\pi<\theta<\pi$, that

(a) $T_n(\cos\theta)=\cos(n\theta)$

and

(b) $U_n(\cos\theta)=\dfrac{\sin(n+1)\theta}{\sin\theta}$

It can be easily proved that the polynomials $T_n(x)$ and $U_n(x)$ satisfy the following Orthogonality properties:

$$(\alpha)\qquad \int_{-1}^{1} T_m(x)\,T_n(x)\,(1-x^2)^{-1/2}\,dx=\frac{1}{2}\pi\,\delta_{mn}$$

and

$$(\beta)\qquad \int_{-1}^{1} U_m(x)\,U_n(x)\,(1-x^2)^{1/2}\,dx=\frac{1}{2}\pi\,\delta_{mn}$$

The generating functions for the Chebyshev polynomials can be defined by using the following expansion formulae:

$$(\lambda) \qquad (1-t^2)(1-2tx+t^2)^{-1} = 1 + 2\sum_{n=0}^{\infty} T_n(x)\, t^n,$$

$$(|\,t\,|<1,\ |\,x\,|<1)$$

and

$$(\mu) \qquad (1-2\,tx+t^2)^{-1} = \sum_{n=0}^{\infty} U_n(x)\, t^n \ (|\,t\,|<1, |\,x\,|<1).$$

Finally, we list below some more results for these special polynomials, the derivation of which can be done exactly in the manner similar to the ones adapted for the other special functions:

(i*) $T_n(x) = U_n(x) - x\,U_{n-1}(x)$,

(ii*) $(1-x^2)\,U_{n-1}(x) = x\,T_n(x) - T_{n+1}(x)$,

(iii*) $1 + 2\sum_{j=1}^{n} T_j(x)\,T_j(y) = \dfrac{T_{n+1}(x)\,T_n(y) - T_n(x)\,T_{nH}(y)}{(x-y)}$

(iv*) $2\sum_{j=0}^{n} U_j(x)\,U_j(y) = \dfrac{U_{n+1}(x)\,U_n(y) - U_n(x)\,U_{n+1}(y)}{(x-y)}$

(v*) $(1-x^2)\,T'_n(x) - xT'_n(x) + n^2 T_n(x) = 0$

We conclude this section by just stating the Chebyshev polynomials play an important role in approximate solutions of differential and integral equations, which aspect of study is beyond the scope of the present book.

SOME PROBLEMS ON CHEBYSHEV POLYNOMIALS

We list below some problems on Chebyshev Polynomials which can be easily proved by just using the definitions and properties mentioned earlier. These problems are of great importance in Numerical Analysis (see [5]).

1. Show that the first few Chebyshev polynomials of first kind, $T_n(x)$ are as given below:

$$T_0(x) = 1,\ T_1(x) = x,\ T_2(x) = 2x^2 - 1,$$

$$T_3(x) = 4x^3 - 3x,\ T_4(x) = 8x^4 - 8x^2 + 1,$$

$$T_5(x) = 16x^5 - 20\,x^3 + 5x, \text{ etc.}$$

and that the following recurrence relation holds:

$$T_{r+1}(x) = 2xT_r(x) - T_{r-1}(x)\ (r \geq 1).$$

2. By reversing the relations in problem 1 above, show that

$$1 = T_0(x),\ x = T_1(x),\ x^2 = \frac{1}{2}[T_0(x) + T_2(x)],$$

$$x^3 = \frac{1}{4}[3T_1(x) + T_3(x)],\ x^4 = \frac{1}{8}[3T_0(x) + 4\,T_2(x) + T_4(x)],$$

and

$$x^5 = \frac{1}{16}[10T_1(x) + 5T_3(x) + T_5(x)].$$

3. Expand $(1 - Pe^{i\theta})$ binomially, for $|p| < 1$, and establish the following result, by taking $\theta = \cos^{-1} x$, and considering only the real parts of $(1 - pe^{i\theta})^{-1}$ and its binomial expansion:

$$\frac{1 - px}{1 - 2px + p^2} = \sum_{r=0}^{\infty} p^r\, T_r(x).$$

4. Establish the following formulae:

$$T_r(x) = \frac{1}{2}[(2x)^r - \{2\,({}^{r-1}c_1) - ({}^{r-2}c_2)\}(2x)^{r-2} + \{2\,({}^{r-2}c_2) - ({}^{r-3}c_2)\}(2x)^{r-4} + \ldots]$$

and

$$U_r(x) = (2\,x)^r - ({}^{r-1}c_1)(2\,x)^{r-2} + ({}^{r-2}c_2)(2x)^{r-4} \ldots$$

[*Hints* : (1) Use the idea of problem 3 and the identity.

$$\sum_{r=0}^{\infty} p^r e^{ir\theta} = (1 - pe^{i\theta})^{-1} = \frac{1 - px + ip\,(1 - x^2)^{1/2}}{1 - 2\,px + p^2},$$

where $\theta = \cos^{-1} x$. Equating imaginary parts of both sides we easily establish a formula, of the type of problem 3, for the polynomials $U_r(x)$.
(2) Write $(1 - 2px + p^2)^{-1} = [1 - p\,(2x - p)]^{-1}$, expand binomially and collect coefficients of p^r of $(1 - px)(1 - 2px + p^2)^{-1}$ and of $p\,(1 - 2px + p^2)^{-1}$, respectively.]

5. Prove that the functions $T_r(x)$ and $U_r(x)$ satisfy the two separate differential equations

(i) $(1 - x^2)\dfrac{d^2 y}{dx^2} - x\dfrac{dy}{dx} + r^2 y = 0$

and

(ii) $(1 - x^2)\dfrac{d^2 y}{dx^2} - 3x\dfrac{dy}{dx} + r\,(r + 2)\,y = 0$

respectively.
[*Hint* : First prove that

$$T_r(x) = A_r (1-x^2)^{1/2} \frac{d^r}{dx^r} \{(1-x^2)^{r-1/2}\}$$

and

$$U_r(x) = B_r (1-x^2)^{-1/2} \frac{d^r}{dx^r} \{(1-x^2)^{r+1/2}\},$$

where A_r and B_r are constants, to be determined, by using the results of problem 4.]

6. Prove that $\int T_r(x)\,dx = \frac{1}{2}\left\{\frac{1}{r+1} T_{r+1}(x) - \frac{1}{r-1} T_{r-1}(x)\right\}$ $r = 2, 3, \ldots$

and

$$\int T_0(x)\,dx = T_1(x), \int T_1(x)\,dx = \frac{1}{4}\{T_0(x) + T_2(x)\}$$

7. Expand an arbitrary function $f(x)$, which is known to be integrable in the interval $(-1, 1)$, in terms of a Chebyshev series $\sum_{r=0}^{\infty} a_r T_r(x)$ and show, by using the results of problem 6, that

$$\int_{-1}^{1} f(x)\,dx = 2\left(a_0 - \frac{1}{1.3} a_2 - \frac{1}{3.5} a_4 - \frac{1}{5.7} a_6 - \ldots\right)$$

References

1. W.W. Bell – "Special functions for scientists and Engineers" – D. Van Nostrand Company Ltd. London, 1968.
2. R.V. Churchill – "Complex Variable and Applications" – McGraw-Hill Book Co., Inc., New York, 1960.
3. E.A. Coddington – "An Introduction to Ordinary Differential Equations" – Prentice Hall, India, 1968.
4. M.S.P. Eastham – "Theory of Ordinary Differential Equations" – Van Nostrand Reinhold Co., London, 1970.
5. L. Fox and I.B. Parker – "Chebyshev Polynomials in Numerical Analysis" – Oxford University Press London, 1968.
6. E.L. Ince – "Ordinary Differential Equations" – Dover Publications, Inc., New York, 1956.
7. J.C. Jaeger – "An Introduction to Applied Mathematics" – ELBS and Oxford University Press, London, 1951.
8. N.N. Lebedev – "Special functions and their applications" – Prentice Hall, Inc., 1965.
9. Yudell L. Luke – "The Special functions and their applications", Vol. I. & II, – Academic Press, 1969.
10. T.M. MacRobert – "Functions of a Complex Variable" – Macmillan, London. 1954.
11. Robert K. Ritt – "Fourier Series" – McGraw-Hill Book Company, New York, 1970.
12. I.N. Sneddon – "Special functions of Mathematical Physics and Chemistry – 3rd Edition." – Longman (London and New York) 1980.
13. I.N. Sneddon – "The use of integral Transforms" – Tata McGraw-Hill Publishing Co. Ltd., New Delhi, 1974.
14. I.N. Sneddon – "Elements of partial differential equations" – McGraw-Hill Book Co. Inc., 1957.
15. G.N. Watson – "A treatise on the theory of Bessel Functions" – Cambridge University Press, Cambridge, 1944,
16. W.E. Williams – "Partial Differential equations" – Oxford, Clarenden, 1980.

Index